The Death of the Dark Energy Idea

in light of ignored failures and inconvenient facts

Subtitle:
The relationship between gravity and quantum mechanics

(The black & white 1st edition.)

By Terrance J. Fidler

The Death of the Dark Energy Idea

Copyright © 2020 Terrance John Fidler
All rights reserved.
ISBN: 9798587439641

Version Update 2026-02-15
Based on a previous unpublished work:
The Dynamic Nature of the Fabric of Space
Copyright © 1990 by Terrance John Fidler
Registration Number TXu 444 524
1990-11-02

Make decisions

Strategy is not just a thought; it is also about completing a task. So once you start generating ideas and making connections between them, you'll have to start making decisions about what to do next. And since there are always limitations in time, money, and resources, you need to learn to prioritize.

When we think strategically, we make unpopular, difficult decisions. To optimize profits, you will have to fire someone or close some product lines. Or what's even more difficult: spend less time on your family to achieve business success. Or vice versa.

The worst thing that can happen in this situation is to fail to make a clear decision and choose the wrong path.

Move away and get closer

We often don't see the forest for a couple of trees. Therefore, a balance is necessary: you need to shift the focus of attention either to the details or to the big picture.

Both approaches have their advantages and disadvantages. When we look at a situation in general, we may miss the most important details and not notice the nuances. If we go into too much detail, we can go in the wrong direction and do something completely different from what we originally planned.

Knowing when to move closer and when to move away is a subtle skill that can only be developed with experience. For now, it is important to know that there are two levels of scale in any situation.

Ask questions

We have already talked about the importance of questions. They help to sharpen things, find out the causes, notice the consequences, and focus on what's important.

Ask yourself questions like these:

1. What in my strategy works and what doesn't?
2. What is changing in the world? In what areas?
3. What do people need?
4. What do I need?
5. How do I know I'm not mistaken?
6. What areas do I need to improve in?
7. What books are worth reading?
8. What are the weaknesses of my strategy?
9. How can they be turned into advantages?
10. What could go wrong?
11. Is this idea rational?
12. Can I trust this source?

Finding answers matters. It turns the brain on full blast and forces it to search for answers. It is important not to stop and throw more and more new questions into your mind.

Put yourself in the other person's shoes

The strategy you created certainly relies on other people as well. This means that if you misunderstand the characteristics of human nature, you can make a lot of mistakes.

You may think that you have built a cohesive team of professionals who will create a new product in the next six months. But this is not enough; it is also important to be able to arouse enthusiasm and motivation in them every day, look for the right incentives, and wisely direct their work.

Learn to resolve conflicts, because without it, your entire strategy may be worth no more than the paper it's printed on. This is incredibly important in business.

Be aware of your own biases

To become a strategist, you need to constantly question your thoughts. Admitting that ideas may be wrong does not affect your authority, but rather the opposite. You are open to testing facts and your thinking, which only allows the mind to evolve.

Ask yourself the following questions:

1. What are the current circumstances?
2. Is my point of view correct? What are its flaws?
3. What influences my thinking?
4. What past experiences led me to this point of view?

Learn to understand the consequences

Every choice has consequences. Once you've created multiple versions of your strategy, think about the consequences of each option. This step is important for making the final decision; It will get easier with practice.

Identifying the effects of different scenarios is important for making the final decision. Ask these questions to evaluate which outcome will best fit your vision:

1. What are the pros and cons of each option?
2. What does each of them mean?
3. Which strategy will best help you achieve your goals?
4. Does this option have a chance to realize long-term opportunities?

DEDICATION

To my grandmother Aune Lemmikki [Korhonen, Ojala] Suni (1930-10-04 to 2016-04-25), who wished she could have had the chance to study more science in school. She was the matriarch of our family and provided stability and someone to look up to for most of us.

Sid Deutsch (1918-09-19 to 2011-01-08) whose work put me on the right path to complete the book. I originally was hoping to partner up with Sid, but shortly after emailing him he passed away.

Halton Christian Arp (1927-03-21 to 2013-12-28) an astronomer of the U.S.A. His work showed that there is an unusually high number of galaxies that seem to have pairs of quasars associated with them. Despite their contradictory redshifts, their proximity to one another seems to indicate that something is wrong with modern theories.

My wife, B.B. Fidler who gave me the time and motivation to work on this. To know thyself.

And a special thanks to Jeff Bezos and his company Amazon, and its division of Kindle Direct Publishing, which allows authors to publish their own work. Without them, I would not have found a means of publishing and distributing my work.

Primary Patron/Sponsor
TBD
Others listed after "About the Author"

CONTENT

	Acknowledgments	i
	Introduction	ix
1	Transverse Waves: The Achilles' Heel of the Big Bang	Pg 1
	An aether / fabric of space versus a multitude of fields and particles	Pg 26
2	Empirical measurements	Pg 38
	Gedankenexperiment	Pg 59
	Sid Deutsch, Louis de Broglie and the Aether	Pg 70
3	The Electron	Pg 77
	The Proton/Positron and Mass	Pg 86
	Gravity, Charge & Protons	Pg 91
	Einstein's Principle of Equivalence	Pg 112
4	The Neutron and the Nucleus	Pg 118
	Particle Physics	Pg 134
	The Anti-Proton and Pions	Pg 138
	The Standard Model vs The Fundamental Model	Pg 143
5	Magnetism	Pg 146
	Magnetism in Two Current Carrying Wires	Pg 162
6	Photons & Electrons	Pg 168
	Max Planck's Quantum Hypothesis	Pg 170
	The Photoelectric Effect	Pg 177
7	The Electric Universe	Pg 198
8	The Cosmic Fog, Redshift & Dead Stars	Pg 205
9	Of Time & Space	Pg 212
10	Conclusion	Pg 217
	About the Author	Pg 218
	Patrons & Sponsors	Pg 219
	Index	Pg 220
	List of Figures	Pg 222
	List of Tables	Pg 226

ACKNOWLEDGMENTS

The following would not be possible without all the contributions and explorations by other people over the centuries. Those who inspired me and those who gave me hints and what I should consider are numerous. The following list does not include the several websites that contained valuable data, and ways to present that data. The future second edition of the 1st edition of this book will contain a list of them.

Present day (2020):
Alexander G. Unzicker – physicist, The Higgs Fake, Bankrupting Physics
Anthony Peratt – physicist, electrical engineer, galactic modelling
David Lindley – science writer, The End of Physics
David Talbot – The Electric Universe
Donald E. Scott – electrical engineer, The Electric Sky
Duncan W. Shaw - author
Emmanuel Fort - silicone oil droplet experiments and pilot waves
Eric J Lerner – plasma researcher, The Big Bang Never Happened
Geraint F. Lewis - The Cosmic Revolutionary's Handbook
Hans Mes – pion research
Henry A. Boorse – professor of physics, The Atomic Scientists
Ignazio Ciufolini – physicist, frame dragging around the Earth
Jacques Hébert – pion research
James Rich - Fundamentals In Nuclear Physics
Jean-Louis Basdevant - Fundamentals In Nuclear Physics
Jefferson H. Weaver – science writer, The Atomic Scientists
Jerome I. Friedman - physicist
Jim Baggott – science writer, Farewell to Reality
John Gribbin – science writer, and astrophysicist
Laura Mersini-Houghton – theoretical physicist
Lee Smolin – The Trouble with Physics
Luke A. Barnes - The Cosmic Revolutionary's Handbook
Margaret R. Boshek – water wave mechanics
Margaret J. Geller – astrophysicist, mapped the Great Wall of galaxies
Margaret Wertheim – science writer, Physics on the Fringe
Mario Gaspero – pion research
Michel Spiro - Fundamentals In Nuclear Physics
Mike N. Keas - Systematizing the Theoretical Virtues
Murray Gell-Mann – quark theorist
Peter Higgs – Higgs field, mechanism, particle
Peter Woit – Not Even Wrong
Ray Gallucci – nuclear engineer, Strobe Star or Neutron Star?
Reinhold Bertlmann – physicist

Segre – antimatter researcher / particle physicist
Sheilla Jones – Quantum Ten, Bankrupting Physics
Shinsuke Nakayama - photonuclear reaction data
Toshiyuki Shizuma - photonuclear reaction data
Tsung-Dao Lee - parity non-conservation of weak interaction
Wallace William Thornhill - The Electric Universe
Yang Chen-Ning - parity non-conservation of weak interaction
Yves Couder - silicone oil droplet experiments and pilot waves

Historical:
Albert Einstein – photoelectric effect, special & general relativity
Alexander Friedmann – cosmologist, expanding universe
Arthur Holly Compton – physicist, the Compton effect
Boris Podolsky – EPR Paradox
Carl Sagan – astronomer, cosmologist, astrophysicist
Cecil Frank Powell - discovered the pi-meson with colleagues
César Lattes - discovered the pi-meson with colleagues
Chien-Shiung Wu - experimental physicist, beta decay experiments
Christiaan Huygens – wave theory of light, 2nd law of motion
Clinton Davisson – electron diffraction
David Bohm - scientist
Edwin Hubble – astronomer, recorded redshift of galaxies
Erwin Schrödinger - physicist
Fritz Zwicky - astronomer
Galileo Galilei – natural philosopher, first to see moons of Jupiter
Georges Lemaître – expanding universe, big bang proponent
Giuseppe Occhialini - discovered the pi-meson with colleagues
Halton Arp – "Seeing Red", "Quasars, Redshifts, and Controversies"
Hannes Alfvén – plasma physicist, Cosmical Electrodynamics
Henry W. Kendall – particle physicist
Hideki Yukawa – pion as the source of the strong force
Isaac Newton – natural philosopher, gravitational equations
James Clerk Maxell – electromagnetic theory of light
Johannes Kepler – astronomer mathematician, elliptical orbits
John Bell - physicist
Karl Popper – philosopher who proposed empirical falsification
Kristian Birkeland – currents for the Auroras Borealis and Australis
Lester Germer – physicist, wave particle duality of matter
Lloyd Motz – astronomer, The Atomic Scientists
Louis de Broglie – matter waves, pilot waves
Marie Skłodowska Curie – radioactivity
Mark W. Zemansky - physicist
Max Planck – Planck's constant and quantization in radiation

Michael Faraday – electromagnetic experimentalist
Nathan Rosen – EPR Paradox
Nicolaus Copernicus – sun centered universe
Niels Bohr – physicist, atomic theory
Owen Chamberlain – antimatter researcher / particle physicist
Pierre Curie – radioactivity
René Descartes – corpuscular theory of light
Richard A. Beth – light showing angular momentum
Richard E. Taylor - physicist
Sid Deutsch – electrical engineer, book – Return of the Ether
Stephen Hawking – theoretical physicist, black hole specialist
Thomas Young – wave theory of light, double slit experiment
Tycho Brahe - astronomer
Vera Rubin – astronomer, flat velocity profile of galaxies
Werner Karl Heisenberg – Uncertainty Principle
Wolfgang Pauli – theoretical physicist
Wolfgang Rindler – first to point out the homogeneity problem

INTRODUCTION

There is a growing number of anomalies that have been uncovered in the fields of astronomy and physics that the most accepted theories, about the fundamental nature of the universe, are having a hard time explaining without invoking some unusual entities and physics, and oddly enough, they choose to ignore some people in other fields of science that already have laboratory-based models that are capable of explaining some of these things. Most people in the general public are not even aware of this and other statements in science that are completely false. You have been trained to accept things that are not even accepted by the majority of scientists. The most deceptive of these is String Theory. Most theorists and scientists abandoned it years ago, and have relegated it into the dump heap of lost causes. It has produced no results and made no experimentally verifiable predictions. Part of the problem has been the entertainment value and the production of shows that are pretty much propaganda for these groups, and cheered on by those who like what they believe it implies about the nature of the universe. In order to maintain funding and their egos they must be as squeaky as possible. Remember the old adage that the squeaky wheel gets the grease. In this case, loads of money. However, the word is getting out about the failure of String Theory and there is no science to back them up. To say the writing is on the wall and their days are numbered is kind of a moot point. But this is not what this book is about. Although the material introduced in this book is another nail in an already closed coffin of String theory. I am mentioning string theory because it is evidence, and an example, that just because you are told something by some intellects you should not necessarily believe it. Question everything! Ask for explanations, but more importantly ask them to show you what the empirical evidence is to support their idea. And not what some equation implies outside its valid domain of application. Engineering 101 - equations are only valid under specific conditions.

What I do like about string theory is that even though it has failed, it has shown that they would like to find the fundamental "stuff" that everything is made out of. Part of their problem was that although they tried to find "it" their stuff would still not explain how things worked. They just were trying to push the problem out once more to a deeper level and said: "This particle will generate the [insert any force name here] force." But they were never going to tell you how the particles actually generated any of the forces. They can't.

Those who have written about the problems with String Theory:
Jim Baggot (Baggot, 2013) – Farewell to Reality
Lee Smolin (Smolin, 2006) – The Trouble with Physics
Peter Woit (Woit, 2006) – Not Even Wrong

What this book is related to but not initially about is our understanding of the role that electricity plays in the universe.

It has helped to motivate me that there are some main stream educators, engineers, physicists and scientists that are not happy with the status quo.

Some of those educators, engineers, physicists and other scientists:
Alexander Unzicker – "The Higgs Fake", "Bankrupting Physics"
Halton Arp – "Seeing Red", "Catalogue of Discordant Redshift Associations", "Quasars, Redshifts, and Controversies"
Hannes Alfvén [Nobel Prize winner] – Cosmical Electrodynamics
Eric J Lerner – The Big Bang Never Happened
David Talbott - The Electric Universe
Donald E. Scott – The Electric Sky
Tom Findlay – A Beginner's View of Our Electric Universe
Wallace Thornhill - The Electric Universe website
And a few others.

Evidence! What evidence do you/we have for this model/theory?
What can your model/theory explain that others cannot?
These are the most fundamentally important questions.

I have tried to let the data and experiments guide me to an answer. With no preconceived notion of what I would find, but that it should point to some kind of a model that actually explains how the forces work, and not just push off the explanation to another hidden particle or force that generates them. I was seeking an actual explanation using physics, backed up by experiments, as to how the forces work to accomplish what they do. For example, how or why are two masses drawn to each other? Current models push off the answer such that we might as well just say two pixies fairies push the two objects towards each other.

Failures of some of the current models:
1. The quark model does not predict radioactivity.
2. The quark model does not predict the negative charge profile measured around the neutron.
3. The quark model does not predict the half-life of neutrons, nor that fact that it has one.
4. The quark model does not predict the existence of ultra-cold neutrons.
5. The quark model requires nine versions each of protons and neutrons. Based on their required quantum chromodynamics.
6. Quantum mechanics cannot explain gravity.
7. General relativity cannot account for quantum mechanics.

The most important error of all is the fundamental failure of modeling photons as transverse waves. This is the most significant error clouding our judgement on some key concepts in cosmology and atomic structure. Extraordinary claims require extraordinary evidence. Or so we are repeatedly told. Even some valid evidence would be nice for the basis of what is currently the most influential theory regarding the structure of photons. Application of transverse wave theory to the most fundamental basis of how photons should work has failed repeatedly, and the "mysterious" data gave rise to the concept of quantization and quantum mechanics. [Quantum mechanics is simply based on localized gradient fluctuations and layers or pockets of gradient formations.] Critically important to all of this is that there is no mechanism of propagation for transverse waves through empty and/or uniform space. Transverse

The Death of the Dark Energy Idea

waves are boundary distortions which require a restorative force to bring back the distortion to an equilibrium position. For bodies of water, it is gravity that pulls down the peaks and the buoyancy force that pushes the trough back from where it was displaced. For guitar strings, it is the molecular cohesion between atoms. There is no infinite set of boundaries permeating all of space for transverse wave photons to be able to travel in any direction. Such complexity is ludicrous.

The alternative of a non-empty space gives rise to using longitudinal density waves which is the basis of a new model that can be used to explain most of everything including the failures of transverse wave theory, like the UV Catastrophe and the expectations of the Photoelectric Effect before the experiments were performed, and show exactly what is the fundamental nature of quantum mechanics and gravity. With an elliptical-frontal profile, longitudinal density waves can be polarisable which also better fits with the data with respect to the number of photons passing through polarising filters [Also as the frequency increases it is harder to polarize what is likely more rounded profiled photons.]. It compliments most of what is experimentally verifiable and extending its' mechanical implications eliminates many of the gaps in our understanding of physical processes. It simplifies our understanding of nature and makes the complexity friendlier. In the new model, we extend longitudinal density waves to reanalyze two-photon physics, where two gamma-rays become an electron-positron pair, and a solution reveals an amazingly simple model that can explain a surprising amount that previously was considered almost magic. Too many people want to believe in such things as the multiverse and travel by wormholes. These ideas sound exciting and interesting, but there is no evidence. A number of unverified implications like wormholes have arisen due to the application of values to equations outside their domain of validity. One of the first principles of engineering is that equations are only valid within a specific domain of use. Valid under only certain conditions.

Let the data dictate the solution. Don't try and cherry-pick the data to support a hypothesis that only works outside a valid domain of an equation. For instance, the equation $I = V/R$ to calculate the current I as the resistance R drops to zero does not ever give a circuit infinite current. Instead, what happens is that all available electrons in the system can flow freely. Applying the equation to a circuit cannot bring into existence electrons that do not exist in a system to give us infinite current. Yet many people apply equations outside their scope of validity.

The Death of the Dark Energy Idea

Starting in 1887 Michelson and Morley, of the famous interferometer light experiments, tried to detect an aether wind flowing over a solid Earth, composed of solid atoms, as it passed through space. But they could not detect any significant changes in the speed of light that made any sense. Yet, we don't talk about a change in the speed of sound as a train passes by us, only a change in frequency. This is exactly what we see relative to the stars around us with red and blue shifts in the frequency of the photons depending on the relative motion of the stars in question. In 1905 Einstein stated that we simply had to accept that the speed of light is a constant and move on. But it was not until 1911 that Rutherford's team showed us that atoms were anything but solid, and in fact were almost entirely composed of empty space. There could be no surface wind, and the detection of the relative motion with respect to other stars by frequency changes just as easily implies longitudinal waves. Longitudinal waves can travel through uniform space, unlike transverse waves and both transverse waves and longitudinal waves are of course limited in speed to their conductor, and thereby giving us a natural explanation for the constant nature of the speed of light. But this is just the tip of the iceberg of what this all implies and leads to.

The Death of the Dark Energy Idea

Chapter One

Of course, we might be wrong but...

The great tragedy of science – the slaying of a beautiful hypothesis by an ugly fact. - Thomas H. Huxley (1825-1895)

TRANSVERSE WAVES: THE ACHILLES' HEEL OF THE BIG BANG

A fatal flaw in the most fundamental model of nature has been overlooked for so long that many people do not question it. The transverse wave model of photons, which are the basic constituents of light, has failed at almost every prediction its proponents made. And because of this reliance on the transverse wave model for photons, this is the Achilles' heel of contemporary Big Bang cosmology. Why? The champions of the Big Bang ideology not only adhere to the basic principles of the waveform, but they also invoke every limitation of the simplified physics of this waveform as some form of an attempt at presenting the purity of their reasoning and, thus, the validity of their conclusions. Despite the accepted probabilistic, messy nature of reality implied by quantum physics.

Extraordinary claims demand evidence, not excuses or waving off the explanation to another dimension or particle. Dark energy is based on the apparent redshift of most of the visible universe, and since transverse wave photons have no normal apparent mechanism to become redshifted, the conclusion made was that the entire universe must be expanding to trigger the observed redshift. And Dark Energy must be the cause. As if, somehow, this astounding explanation seemed reasonable! Even though experiments based on the transverse wave photon model failed, like the photoelectric effect, and that longitudinal density waves behave in a similar manner in experiments.

Although Albert Einstein helped lay the foundations of quantum mechanics through his work on the photoelectric effect and the concept of light quanta (photons), he remained deeply uneasy by the work of others implying the theory's apparent assertion of randomness. This discomfort is often cited as the reason behind his famous remark: "God does not play dice with the universe." Einstein's comment was directed at Max Born's interpretation of quantum mechanics, which suggested that chance plays a fundamental role at least at the atomic level.

Against this backdrop, Einstein sought a model of the universe that preserved rationality and determinism. He turned his attention to possible mechanisms underlying the conduction of photons. In doing so, he proposed what were in effect "pilot waves," later sometimes called "ghost fields." The phrase "ghost fields" (Gespensterfelder) is often attributed to Erwin Schrödinger, who used it in a somewhat mocking tone. Like Einstein, Schrödinger was also uneasy with the implications of quantum mechanics, and his remark highlighted the elusive and mysterious qualities of such pilot-wave ideas.

From a mechanical perspective, only longitudinal compression waves naturally exhibit a leading, gradually compressing region that precedes the main high-density wavefront. This frontal region can be understood as the pilot wave that at a minimum precedes the main body of the photon. By contrast, no such region can exist if the photon is modeled as a purely transverse wave.

An illustrative example of transverse waves is the undulating motion observed in ocean waves, representing a traveling boundary-wave distortion between two dissimilar materials, such as water and air. Classically, transverse waves are defined as oscillations occurring perpendicular to the direction of the wave's propagation. In other words, the waves peak in height vertically while moving horizontally.

Today's prevailing cosmological model, depicting the universe's appearance, is built upon the concept of the Big Bang, a primordial explosion, and the supporting evidence for this model relies on the transverse wave model of photons. However, this approach overlooks a fundamental scientific principle – the necessity of establishing a solid foundation for any model. Ignoring inconvenient truths undermines the credibility of the starting point for any scientific model. The widely accepted understanding of the universe's underlying nature hinges entirely on the model of photons as transverse waves. Yet, a critical question remains unanswered: how do transverse waves propagate through empty space, and what restoring force triggers their propagation?

In contrast to the restorative forces at play in water waves driven largely by gravity, the proponents of the transverse wave model still need to explain the mechanism through which these waves travel through space. The photons' transverse wave model's adherents expect acceptance without questioning its operational principles, asserting that photons pass through space. However, two observational facts that support the transverse wave model exhibit fatal flaws. Additionally, two critical experimental failures further undermine the coherence of the model.

An alternative approach considered modeling photons as longitudinal waves. However, this was troublesome because it required the existence of the luminiferous aether, a universal medium postulated to propagate light waves as longitudinal compression waves. But previous attempts to both detect and propose how energy and matter interacted with such a medium eluded investigators of the past. In my proposed model, I refer to this medium as the Fabric of Space (FOS). The famous Michelson & Morley experiments, initiated in 1887, were intended to prove or disprove the existence of the aether. Their well-known c + v experiments, as shown in Figure 1, were designed to detect whether the Earth's motion (v) affects the observed velocity of light (c), by measuring shifts in the interference fringe pattern of the light spectrum. This method relied on the principle that if light traveled through a stationary aether, then the Earth's motion through it would cause the speed of light to vary with direction — a phenomenon known as anisotropy — which would appear as a shift in the interference fringe pattern observed with an interferometer. The absence of such a shift was historically interpreted as evidence against the existence of aether. However, alternative interpretations suggest that the null result may instead reflect properties of a compressible or entrained medium, rather than its nonexistence.

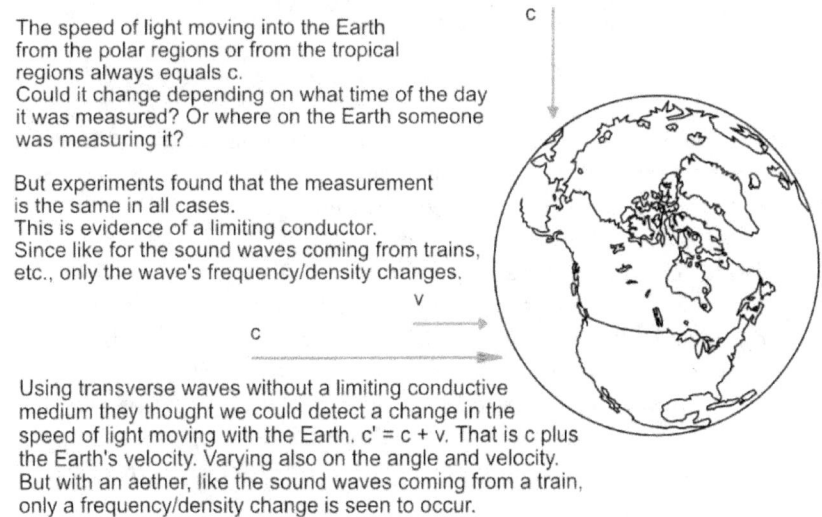

Figure 1 - c+v for the Michelson and Morley experiments

Figure one shows the indicators of light as it approaches the Earth from both a polar pole, and the tropical regions at right angles with respect to these regions. Michelson and Morley tried to see if they could detect a change depending on what time of day it was measured, what season it was, or where on the Earth they attempted to measure it. But the experiments found that the speed of light did not vary no matter how, where or when they measured it. This is evidence of a wave limiting conductor. Since like sound waves coming from trains, etc., only the frequency/density can change. Not the velocity of the wave.

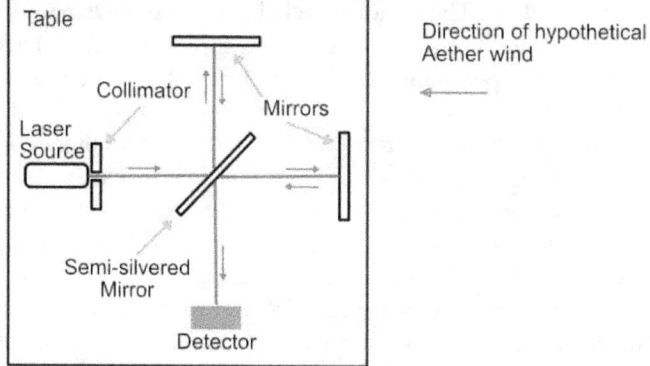

*Note: All of these ideas were developed at a time when the Earth, all matter, was believed to be truly solid atomic atoms. The experiments started in 1887 but not until 1911 did the Rutherford experiments showed that matter is composed mostly of empty space. ~ 99.9999999999996% 'nothing'

Figure 2 Michelson and Morley experimental setup

Figure two shows a simplified version of the instrument that Michelson and Morley used to try and detect a change in the velocity of light, based on the changes in the interference patterns that would be generated.

This idea of adding any additional component to the speed of light, c, is critically flawed because first, they are treating photons as particles independent from their conductive medium. In the Doppler Effect for moving things like trains, sound waves only change frequency, not velocity, when the source generating them moves. The Doppler Effect is the same for the photons coming from other stars. Stars moving towards us have a blueshift in their spectral signatures, and a redshift when they are moving away from us.

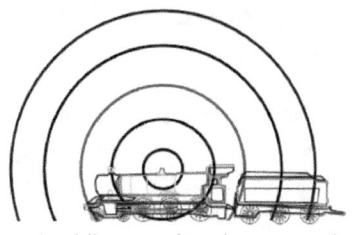 

The wavelength/frequency of sound waves emanating from a stationary source are the same in all directions.

While those coming from a moving source are compressed ahead of it and stretched out behind it. Although if an observer moves in unison with it the observer sees no such change.

Figure 3 - The Doppler effect for sound

Figure three shows the standard comparison of sound from a stationary train and that the frequency that is being generated is uniform in all directions, while for a moving train that the frequency changes measured, or experienced, depends on whether the train is moving towards or away from you.

Instead, we should be looking at the density changes which are equivalent to frequency changes. So, instead of c+v we, and trains, are actually working with normal density plus or minus some density change, $+/-\Delta d$, which corresponds to a change in frequency.

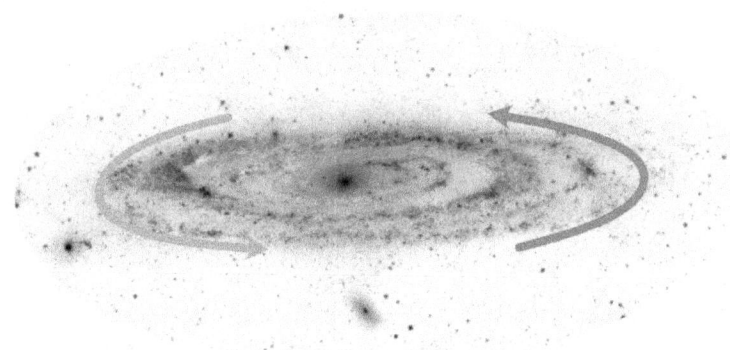

The Andromeda galaxy is a perfect example for simultaneously observing the blue-shift and red-shift of photons. The part of the galaxy rotating towards us is blue-shifted with respect to its center. While the part moving away from us is red-shifted with respect to its center.

The Doppler effect is the result of the relative change in the compression or decompression of the body of the photons, just as it is for sound waves, as seen by an observer. Where an increase in density is seen as being blue-shifted which we interpret as an increase in frequency and a decrease in wavelength. While a decrease in density is seen as a red-shift and thus a decrease in frequency but an increase in wavelength.

Figure 4- The Doppler effect for light [Credit: NASA image of M31]

Figure four shows an image of the Andromeda galaxy, a NASA image, and that by looking at the redshift and blueshift of its' stars shows that there is a blueshift of the stars on the left side of the image while the right side of the galaxy produces a redshift in their frequencies. Thus, showing the galaxy is rotating counter clockwise with respect to us.

An increase in density is seen as being blue-shifted which we interpret as an increase in frequency and a decrease in wavelength. While a decrease in density is seen as a red-shift and thus a decrease in frequency but an increase in wavelength.

*Note: Not to scale.

Figure 5- Hydrogen lines of the Balmer electron orbital transition series

Figure five shows how the blue and red shifts are actually determined. By comparing the Hydrogen absorption lines with respect to a non-moving laboratory source. If the absorption lines shift towards the red then the star is determined to be moving away from us. And vice versa.

The proponents of the transverse wave model are claiming that the historical experimental evidence irrefutably supports photons as transverse waves, and that they cannot be longitudinal compressional density waves. Similar in nature to sound waves – phonons. There are not only critical flaws with their line of reasoning, and what they believe to be their evidence, but what is even more problematic is their blatant disregard of the failures of the transverse-wave photon model. While simultaneously ignoring the evidence for photons as longitudinal compressional-density waves.

The main problem, an intuitively preposterous notion, that has arisen has been some of the implications of the redshift of photons based on the transverse-wave model. The required cause of the apparent increasing redshift of photons from galaxies further and further away from us, has inevitably led to some of the most exotic and astounding lines of reasoning about the past and future nature of the universe. It has led to the Big Bang model from the apparent expanding universe, and with some of the most distant galaxies with velocities approaching the speed of light. A physics problem in itself. The physical limitations of transverse waves has inevitably led to this, as they have a fixed way to behave and decay and the conclusions based on this are logical even if they are astounding. And yet almost everyone has chosen, out of the desire of wanting a big bang in the past, to ignore the restoring force, of the conducting medium, required for transverse waves to propagate through space. Along with failing to review the basic principles of the experiments with water transverse-waves as they pass through the gaps of barriers and around objects. Experiments which were performed by classical physicists of the past. They seemed to have reported what they observed, but failed to investigate what was causing the collapse of the peaks and troughs of the water transverse-waves as they passed through gaps in barriers, or around objects. Their reverence for what they were taught, show that those who follow down this path, of needing transverse waves to represent photons, have no interest in finding errors in their reasoning and triggered their inevitable inability to see what was happening in front of them.

The Big Bang that most cosmologists demand, absolutely requires transverse-wave photons to obey the rules/properties of their fundamental nature. We are told the redshift of transverse waves can only be achieved by some mechanism of the most distant galaxies being forced away from us and/or the universe undergoing a rapid expansion. Dark energy is believed to be the source of that energy. The further away a galaxy is, the greater its velocity appears to be. That is their red-shift appears greater. Yet roughly speaking the distribution* of galaxies appears to be uniform everywhere in contradiction to what the model of the expanding universe predicts, and then there is the relative separation* velocities that show no difference between galaxies that are relatively close to one another whether or not those galaxies are closer to us or out at the edge of the observable universe. This unexpected uniformity is known as the Horizon problem. Wolfgang Rindler pointed this homogeneity problem out in 1956.

The Horizon problem, which is derived from galaxy distribution surveys, and data from the Cosmic Microwave Background Radiation (CMBR) surveys, shows that the galaxies appear to be uniformly distributed. Yet, we are told to believe in the big bang expansion despite the fact that just like our neighboring galaxies appear to be moving relatively slowly in comparison to us, so to are the most distant observable neighboring galaxies in relationship to one another. Add to that the constant increase in velocity outwards based on the observed redshift, such that the most distant detectable galaxies some 13 plus billion light-years away, appear to be approaching the speed of light, and still the distribution of the galaxies out there is the same as it is closer to us. By definition if you have the same amount of material occupying greater and greater volumes, this should result in a lower density. There should be greater distances between galaxies the further out they are. Ignoring of course the normal standard group formations. Why is this so important with respect to the most accepted cosmological model that it is based on transverse waves? Because only transverse waves are supposed to be polarizable, and only transverse water-waves have been observed to spread out after passing through a gap or around barriers. This is wrong on both accounts.

We are told that the "simplest" explanation, for the redshift, is that the entire universe must be expanding because only transverse waves, that show no sign of some sort of energy decay/absorption mechanism, can be shifted to the red part of the electromagnetic spectrum by such an extreme amount. No mechanism or other data supports them increasing their wavelength by any other means. Seriously?! Expand the entire universe, or that the bodies of longitudinal-like photons simply physically spread out/expand in part due to the limitations of their mechanism of propagation? Which sounds more plausible?

So, if the fundamental property of the physics related to frequency changes, based on the light emanating from distant stars which appear to be traveling at higher and higher velocities the further out they are, can be invoked to allow transverse waves to undergo a redshift - is the key evidence of the big bang, how is it that the even most basic property of their propagation can be ignored? Transverse waves require a restoring force to bring back such waves to the equilibrium position required for them to propagate along some material boundary. For bodies of water, it is gravity that pulls down the peaks and the buoyancy force that pushes the trough back from where it was displaced. For large waves it is gravity, but for small waves it can be just surface tension. For guitar strings, it is the molecular cohesion between atoms. Conduction of a transverse wave by definition is not possible in a uniform 3D space. The all-pervading Higgs field does not qualify as such a conductor either for transverse waves.

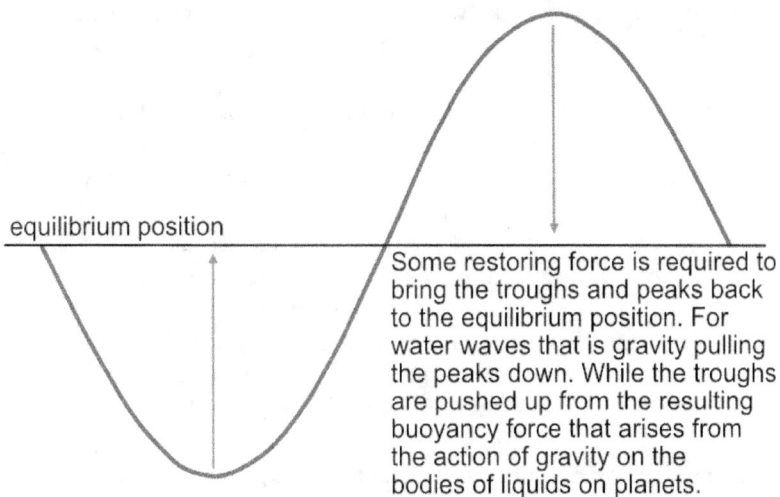

Figure 6 - return to the equilibrium position requires a restoring force

Figure six shows a typical sine wave cycle of a transverse wave. In order for a transverse wave to propagate the peaks and valleys, that are the physical structure of the wave, a restoring force has to exist, or the wave could not exist at all. For water waves on large bodies of water that is gravity pulling down and the buoyancy force, triggered by gravitation on the liquid body, pushing the troughs back up.

Is there any other data supporting the problem with using transverse waves as the model for a photon? Hell yes!

Application of classical transverse wave characteristics to the experimental material world completely failed for two critical phenomena that affect almost everything we consider important:

1. The Photoelectric Effect is about the ejection of electrons from the surface of metal plates, and the application of transverse wave rules completely failed as they predicted that the intensity/ brightness of the light mattered and that it would take some time for it to be triggered. Completely wrong. Only the wavelength mattered, and it happened instantly and started only at some wavelength that depended on the metal being tested. Einstein extended Planck's quanta to photons. This effect is how photons interact with the electrons that are in orbit around some atom, and this interaction is critical in allowing life to exist. A photon that is modeled as a longitudinal compressed density wave easily explains the effect it has with electrons simply depending

on what is the density of the region the electrons occupy, and the density of the photon in question.
2. The Ultraviolet Catastrophe arose because classical physics predicted that if you shined a light on an ideal black body that it would begin to emit radiation in all frequency ranges, as the energy of the peaks and troughs added together, emitting most of that energy in the UV spectrum [x-rays and γ-rays (gamma-rays) were unknown at the time.]. It predicted that a blackbody would release an infinite amount of energy, contradicting observation and the principles of the conservation of energy. From this problem arose the idea of quanta and eventually to our current model of photons. This does not happen because electrons inhabit tiered density layers whose positions are not easily disturbed. [More on this in coming chapters.]

It is the even more fundamental nature of transverse wave propagation that is supposed to be evidence that there can be NO aether-like material in space, as space would have to consist of some bizarre solid, due to the transverse wave requirements, and yet allow the particles from the standard model to pass freely through it. Now the Higgs field is taking on the role of something filling all of space. Ironically its mechanism of crowding and thereby triggering a mass-effect is almost identical in nature to how an aether mass-effect is generated in the Fabric-Of-Space model introduced in this book.

Why is there a conflict in choosing to embrace all the properties of transverse waves completely? And instead, ignore some of the inconvenient basic properties of transverse waves? It is their need to keep the idea of an expanding universe as it is. To keep the Big Bang no matter what the cost. It sounds like epicycles[1] added to epicycles to keep the current model alive, to save face and their jobs. Tired light theories fail because they are still based on transverse waves which in order to experience a drop in frequency, something has to absorb that energy. Or at least disperse it like the weakening waves emanating from the point where a pebble has been dropped into a pool of water. There is both a geometric dispersion factor as well as a frictional component with water waves. Halton Arp showed that energy loss due to interaction with gases is not supported by the evidence because there is no difference in redshift between similarly distant regions that have gas clouds between them and us, compared to those that have none or very little. (Arp 1998) All of this has arisen because transverse waves were supposed to have the unique feature of being polarizable. Not anymore, I will present to you polarization for longitudinal waves shortly. And as I just pointed out, photons

---

[1] This is a reference to the work of Johannes Kepler who showed that the epicycle model for the planets was wrong. Their paths instead were elliptical. Not the perfect circles added to circles demanded by the theorists in his time.

do not appear to support themselves as being based on transverse waves. The photoelectric effect and UV catastrophe show this clearly. Let alone the lack of any supporting model for the physics of the propagation of transverse waves in a uniform and boundless three-dimensional space. No restoring force can be invoked in the transverse wave model. How many strikes against it does it take for people to acknowledge these key fundamental problems?

In short, and I believe it is appropriate – "The Emperor is not wearing any clothes!"

We have been told that all the data supports the expanding universe. No – only the apparent redshift. Nothing else. Everything else has repeatedly been modified to fit observation, theories have been dramatically altered to adapt when facts were showing them to be way off, and no experiment in a reasonable amount of time can be done to support their conjecture about an expanding universe. Of course, theories need to adapt to the facts, and thus adaptation is expected to be part of the process. However, when you confuse empirical data that supports a specific outcome of an event, with data that does not actually support the first premise then we can confuse the nature of the relationship of the data with observation. Correlation does not imply causation. Astronomers are now detecting objects at the visible edge of the known universe that defy the Big Bang timeline. Stars too rich in more complex nuclei have been discovered at the visible edge of the known universe, galaxies too complex, and objects claimed to be too big for normal black holes. All of which defy our understanding due to the hypothetical too-young age of our universe based on the big bang theory. These problems support that there is something wrong with the foundation of the model, and that ultimately goes back to photons being modeled as transverse waves.

Carl Sagan's comment on Halton Arp's work and the redshift, and a reoccurring pattern of quasar pairs with some apparent physical relationship to galaxies:

"There is nevertheless a nagging suspicion among some astronomers that all may not be right with the deduction, from the red shifts of galaxies via the Doppler effect, that the universe is expanding. The astronomer Halton Arp has found enigmatic and disturbing cases where a galaxy and a quasar, or a pair of galaxies, that are in apparent physical association have very different red shifts. Occasionally there seems to be a bridge of gas and dust and stars connecting them. If the red shift is due to the expansion of the universe, very different red shifts imply very different distances. But two galaxies that are physically connected can hardly also be greatly separated from each other—in some cases by a billion light-years. Skeptics say that the association is purely statistical: that, for example, a nearby bright galaxy and a much more distant quasar, each having very different red shifts and very different speeds of recession, are merely accidentally aligned along the line of sight; that they have no real physical association. Such statistical alignments must happen by chance every now and then. The debate centers on whether the number of coincidences is more than would be expected by chance. Arp points to other cases in which a galaxy with a small red shift is flanked by two quasars of large and almost identical red shift. He believes the quasars are not at cosmological distances but instead are being ejected, left and right, by the "foreground" galaxy; and that the red shifts are the result of some as-yet-unfathomed mechanism. Skeptics argue coincidental alignment and the conventional Hubble-Humason interpretation of the red shift. If Arp is right, the exotic mechanisms proposed to explain the energy source of distant quasars—supernova chain reactions, supermassive black holes and the like—would prove unnecessary. Quasars need not then be very distant. But some other exotic mechanism will be required to explain the red shift. In either case, something very strange is going on in the depths of space."

(Sagan 1985) – from the book "Cosmos"

If all the properties of transverse waves fail to be relevant, or necessary, when it is convenient to ignore them for a theory intended to be used for a model of the real world, then something is clearly wrong with their reasoning and thus obviously their conclusions as well. The ability for transverse waves to be polarizable was one of two key phenomena, the other being the way in which water waves spread out after passing through gaps in barriers and around objects, which was supposed to support them as being the best model for photons. First, they failed to consider that longitudinal waves having a frontal profile that are elliptical, not a perfect circular front, allows them to be polarizable. And second, no one seems to have acknowledged or realized that as waves pass through gaps or around objects that it is gravity pulling down on the sides of the peaks and the sides of the troughs of water waves that explain their spreading out. In other words, the physical distortion in the medium conducting the waves changes the path of the transverse waves. More on this in the next few pages.

The transmission of the longitudinal wave photon is the most efficient known mechanical action in the universe, and yet it is not perfect. Therefore, over time, the material making up the photons spreads out, the photon density is reduced, and their wavelength increases. It is not loss, but is instead just a physical transformation that results in their expanding and spreading out. See the next image for a more detailed explanation of how we can get a transverse wave form from the mapping of the electrical and magnetic components from the electromagnetic spectrum of light, and the corresponding longitudinal waveform from which they can be derived. Redshift cannot be avoided. The apparent redshift of photons has incorrectly been interpreted as their sources moving faster & faster away from us, and closer & closer to the speed of light. With today's most popular theory having entire galaxies approaching the speed of light despite the energy it would require to move such masses in violation of Einstein's work. Thus, the basic premise and evidence for the Big Bang is simply the wrong interpretation of the data because they used the wrong model for photons.

The first half of the following image, the sinusoidal-like waveform, was supposed to be critical in showing that photons must consist of a transverse wave, which ultimately led to its proponents overlooking the obvious, and ultimately led to the failure in their reasoning. Observe that the typical schematic of an electromagnetic wave as a transverse wave, when compared to the longitudinal wave of compression and rarefaction regions, are, in effect, complementary images of one another. The peak maximum density of the longitudinal wave correlates to the maximum of the electric field, while the minimum electric field corresponds to the lowest density phase of the rarefaction region/phase. Note that the image shows just one cycle, but additional cycles can occur, thus extending the wave pattern out due to the recovery process and over-compensation of re-compression. The pressure difference between the surrounding background environment and the compression & rarefaction regions of the photon's body influences magnetic field formation due to the localized motion of the aether moving away from then back towards the disturbance created by the wave passing through. This pressure difference likely also contributes to the photon's rate of growth or expansion.

One of the other key pieces of evidence that photons were supposed to be transverse waves is that the electric field's apparent maximal perpendicular strength was interpreted as the electric field is being perpendicular to the direction of travel. And thus, it was believed that they must be like the transverse waves that travel on the surface water. They had not considered it instead to be the maximum density, and thus equally as well as being in the maximum potential state of doing work.

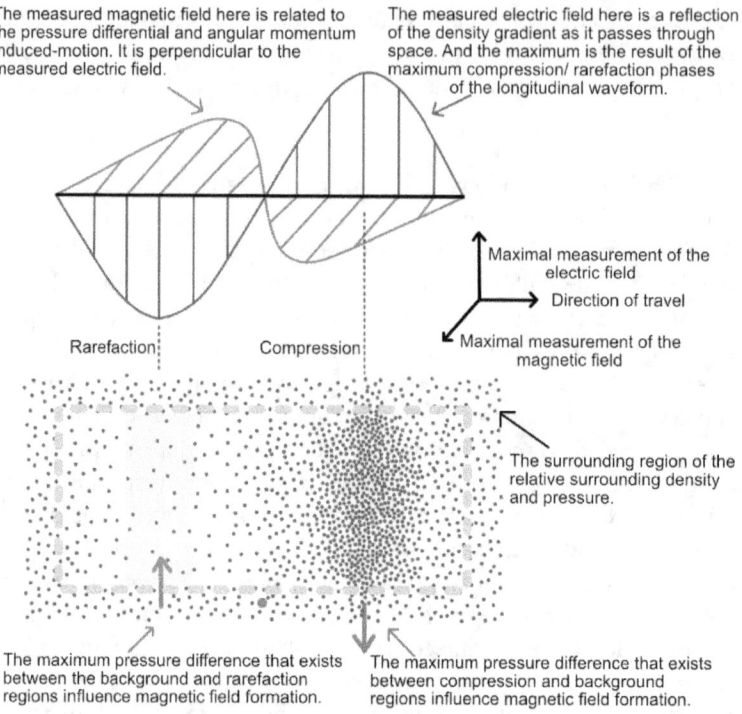

Figure 7-Transverse wave mapped onto a longitudinal wave

Figure seven, perhaps one of the most important concepts in this chapter, is that if you map the density and momentum of a longitudinal wave, based on the existence of some conducting medium, then you end up with what looks like a sine wave-form. Whose form was supposed to be the proof that a light wave has to be a transverse wave.

Longitudinal waves spread out naturally, crudely speaking like the spreading out of the circular pattern of transverse waves generated by a dropped pebble on the surface of a pond. Like them, you cannot keep a longitudinal wave focused so that it does not disperse. The energy and form of the transverse waves emanating from a pebble dropped onto a pond spread out naturally due to the ever-expanding curved wavefront generated by this event. The local potential energy is lost, but the frequency stays the same. This is something that is every rarely commented on in any books.

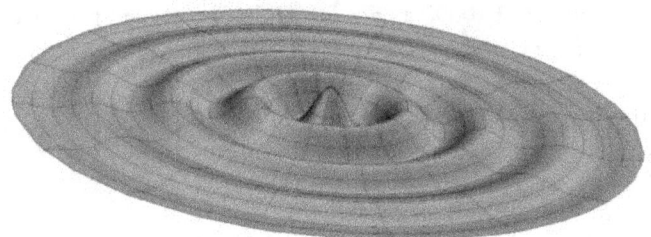

Figure 8 - circular transverse waves spreading out on a pool of water

It is not so much the loss of energy that mostly causes the waves to diminish, but instead the mechanics of an expanding-out circular wavefront. With them getting more and more weaker the further they get away from the center from which they arose. The potential-energy loss is more like the physics related to sound waves spreading out in a gas and their phonon body dropping in density. And thus, we get redshift without the loss implied by tired light theories.

In the following figure portraying a longitudinal based photon, the angle θ is not zero, which is therefore forcing its expansion and thus redshift. This is therefore also the smallest detectable angle in the universe that can be detected from the evidence of red-shifted photons. The volume increases, $V_2 > V_1$, while the density decreases, and thus redshifted photons, like all comparable other photons, have a lower potential to trigger the photoelectric effect.

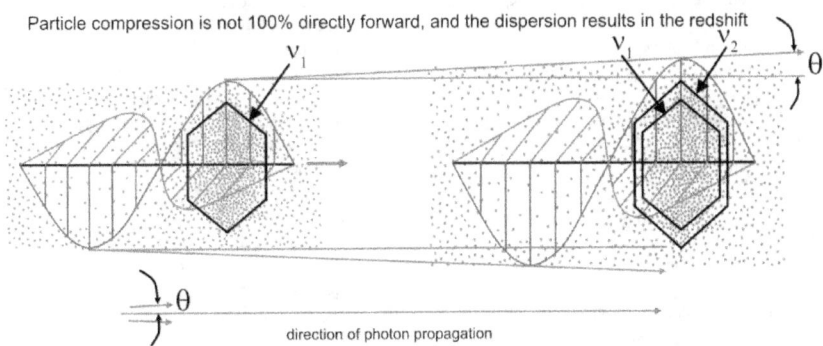

Figure 9 Particle compression is Not 100% directly forward, this triggers redshift

Figure 9 shows that as a longitudinal wave in some gaseous medium is transmitted, the wave enlarges and thus the density drops, and the wavelength increases. A naturally occurring redshift without needing dark energy to trigger the expansion of the entire universe.

Particle compression is not 100% focused directly forward, and add to that the pressure differential that exists with respect to the body of the photon and the medium propagating it, and therefore the volume of the photon increases slightly over lightyears, and this physical change is seen by an observer as the famed redshift. No need to demand that the entire universe must expand to accommodate the defective model using transverse waves for photons.

Johannes Kepler killed the epicycle upon epicycles view of the motion of our planets when he showed that in fact the motion of the planets are elliptical in nature. Not perfect circles as the elite of the day commanded the planets to move.

"Only the most intelligent people can see he is wearing the most spectacular clothes!" It took just one child, unaware of the adult's desire to be part of the trendy collective, to point out the obvious fact that the Emperor was naked.

Think what it would mean if it could be shown that the entire premise of the big bang appears to be wrong! And not only that, if you follow the empirical data that much of what we think of as a mystery is actually easily explained by longitudinal density waves combined with de Broglie's [& Deutsch's] barely detectable, but influential, compression pilot waves. All under the control of a longitudinal wave conductive medium.

Transverse waves	Longitudinal density waves
Extreme redshift due to speed/expansion	Redshift due to aether wave spreading out
Expanding universe	No expanding universe
Quantum Emergence	No quantum emergence
Dark energy	No dark energy
Big bang	No big bang [Mersini-Houghton, plus.]
Antimatter formation during big bang	No antimatter event - did not happen
Planck epoch	No Planck epoch
Reheating era	No reheating era
Particle era	No particle era
Photon era	No photon era
Recombination epoch	No recombination epoch
Electroweak epoch	No electroweak epoch
Inflationary epoch	No inflationary epoch
Grand unification epoch	No grand unification epoch
Quintessence Field	No quintessence field
Cosmic Microwave Background Radiation	CMBR persistence problem

Table 1- Transverse waves versus Longitudinal density waves.

Longitudinal, compression-rarefaction density-varying, waves with an elliptical profile are not only polarizable, but you automatically get a mechanical solution to the photoelectric effect due to density. And using de Broglie's [& Deutsch's] pilot waves that form ahead of photons and particles – the material that represents the most weakly compressed formation ahead of them, and the related density/pressure differentials as they pass through a region of space, we get a mechanical explanation for a guidance mechanism for both the photon and electron two-slit experiments, and photon single-slit and edge-of-object experiments. Pilot waves travelling ahead of both photons and electrons provide a mechanical solution, along with the medium's changes, to these two key quantum phenomena. The fact that sounds waves do experience the same thing, shows that mechanics alone can explain what was once a mysterious quantum effect. Pilot waves also cover the so-called mysterious Observer Effect. The key here is the use of a device that can detect the presence of the passage of a photon through one of the slits. Once close enough to the slit, it interferes with the flow of the pilot waves. The whole idea is that when observed, they do not spread out, but when not observed, they form the normal spread-out distribution pattern. I believe this is solved by the unobserved photons' pilot waves behaving normally, while for the observed ones, where the detection instrument approaches one of the slits, they disrupt the pilot waves normal flow and thus interfere with the usual conducting medium's collapse, or deformation, and spreading out of the pilot waves. And with that, trigger a new pattern for the photons.

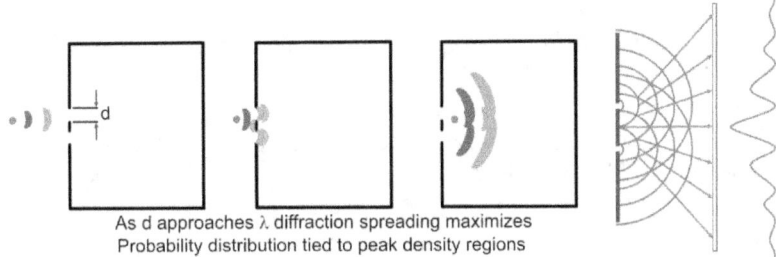

As d approaches λ diffraction spreading maximizes
Probability distribution tied to peak density regions

Figure 10- Two slit experiments for electrons or light

Figure 10 shows that in double slit gap-diffraction that as the pilot wave of the photon approaches the double gap that some energy is reflected back, and that the pilot wave is split to pass through the gaps. Now the two pilot waves that are formed that pass-through overlap in some regions, and it is these doubly dense regions that are believed to be the better conductors and thus the probability of the photons being here increases. Quantum mechanics with a simple physical explanation.

A model for photons, and electrons, with pilot waves is supported by Yves Couder's [et al.] silicone oil droplet experiments, from the University of Paris, showing quantum-like interference for two-slit experiments arising from the "pilot" waves that are generated on the surface of the fluids in the experiments. Similarly, photons and electrons are guided where the peaks of the complimentary pilot waves unite, thereby providing them with a better path of conduction.

Instead of thinking about the photons on their own, in two-slit experiments, think instead of them being preceded by their paired pilot waves which then guide them to the screen in discrete probabilistic intensity patterns. Quantum mechanics is no longer a mystery under this model with such a simple mechanical explanation for the two-slit experiments. The distance between peaks of the recorded pattern is related to the wavelength, or more appropriately the wave spacing between potential photons following each other. While on its own, a photon is affected by the minimum necessary density of its own split pilot waves to trigger guidance by conduction due to the regions of the pilot waves meeting and enhancing each other.

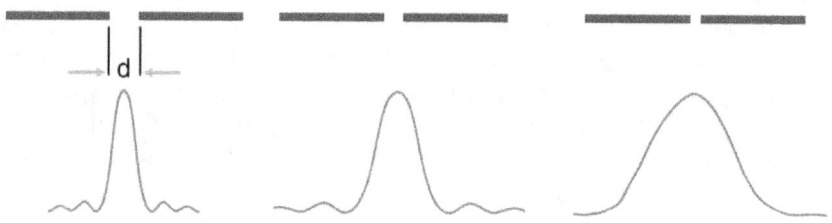

Figure 11- As the gap distance approaches the wavelength spreading maximizes

Figure 11 which shows that as the gap 'd' becomes smaller and thus approaching the wavelength of the photons being used, that the peaks and troughs of probability spread out.

The same is true for the quantum mechanics of the electron orbitals of atoms. This is where de Broglie's work with Einstein was crucial in allowing Erwin Schrödinger to develop what would eventually be called the Schrödinger Equation that gave a wave equation for matter, which was then used to provide us with the equation for predicting the orbital position of electrons around nuclei. But not why it works. Around atoms/nuclei, quantisation is due to tiered density layers that are triggered by electrons as they are guided by the best conductive path ahead of them, with the distinctive layers being formed due to electrons' vortex-like secondary property of spreading out the material they have passed through. Wake dissipation regions are formed. With tiered density layer formation being controlled by the recovery rate of a density region in conjunction with the area-volume associated with its orbit, and with respect to the granularity of the aether implied by Planck's constant. FOS density varying volumes are the probabilistic quantum regions, as is shown by the nature of electron orbitals.

It is not some property of transverse waves that forces the waves to spread out the water after they pass through a gap in breakwater gap diffraction, or the edge of a barrier. This diffraction also happens to longitudinal sound waves. Instead, as the water wave passes through a slit, its higher amplitude peaks find their sides closest to the slit no longer supported and thus gravity pulls down the peaks of the waves around the corner, while alternately the sidewalls of the troughs that pass through are pulled into troughs, and thus the waves spread out. This is by-product of the action of the waves on the surrounding medium triggering unsustainable gradient differentials in the medium that need to be compensated for, and not directly due to the transverse waves themselves.

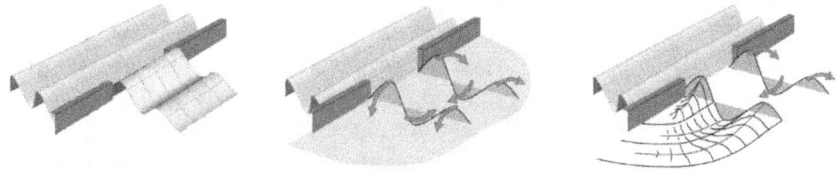

Figure 12-Breakwater gap diffraction of water waves

Figure 12 shows that as water waves passes through a gap in a barrier wall, that it is gravity that pulls down the sides of the peaks, and the walls of the troughs thereby triggering the spread of the transverse wave pattern. This is happening only and because the medium transporting the wave is deforming.

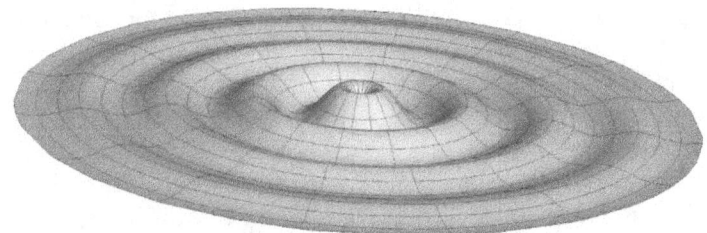

Figure 13- Boundary wave generation

There are a variety of causes that trigger the formation of waves on the surface of water. Events that add energy to the water, or trigger displacement, like the wind, falling glacial ice, landslides, earthquakes, pebbles, etc. These all-trigger conditions that as the system attempts to restore equilibrium, in the supporting medium, the disturbed surface generates waves. Mostly that is gravity, and the buoyancy it triggers, for large waves. At the same time, ripples can be solely supported by the surface tension of water generated by the cohesive attraction of the water molecules to each other. Note that in this "pebble drop" event, in the previous figure, that the waves diminish in amplitude but not frequency. But more importantly, this spreading out changes the density of the energy per unit volume, due to the volume of this "energy" spreading out. Frequency does not play a role in the density. Here only the amplitude can be seen to be correlated to the maximum potential work, lifting up, that the waves can do.

Breakwater gap diffraction actually comes into existence not only for large waves whose sides are clearly affected by gravity, but also for much smaller waves where the restoring force is due to the surface tension of the water due to the molecular cohesion between H_2O molecules. Thus, the spreading out of transverse waves through gaps in barriers, or around objects, is not a property of transverse waves themselves, but in fact due to the properties of the conducting medium and the forces acting on it generated by the degree of the distortion of the medium triggered by transverse waves. In other words, the reaction of the medium past the gap, or object, is due to the magnitude of the disturbance forced upon it by the transverse waves. Thus, allowing the transverse waves to be propagated into regions that were not directly in the path of the original transverse waves.

Transverse waves create the conditions in which gravity can pull down the sides into the troughs and pull down the peaks to the surrounding water level. Through this action, the transverse waves transfer their wave motion along this new path. The transverse waves follow the water's surface as it deforms as it passes through barriers and/or moves around objects. If the side of a tank containing water collapsed, or the glass was removed, the water would spread out directly outward and sideways. Billowing out sideways just the same as water waves moving around an object, but in the absence of any transverse waves. Similarly, as a sound wave passes the edge of some object or doorway, it is the unsupported high-pressure compressed region that all of a sudden finds no support for the compressed material, and so it begins to expand outwards and effectively moves around this corner. See the breakwater gap diffraction figure. Pilot waves provide a similar mechanism for photons and electrons as well. Since their presence would represent a higher density region preceding the photons that then react to the lower pressure region it encounters and spreads out.

Diffraction of Particles vs Waves

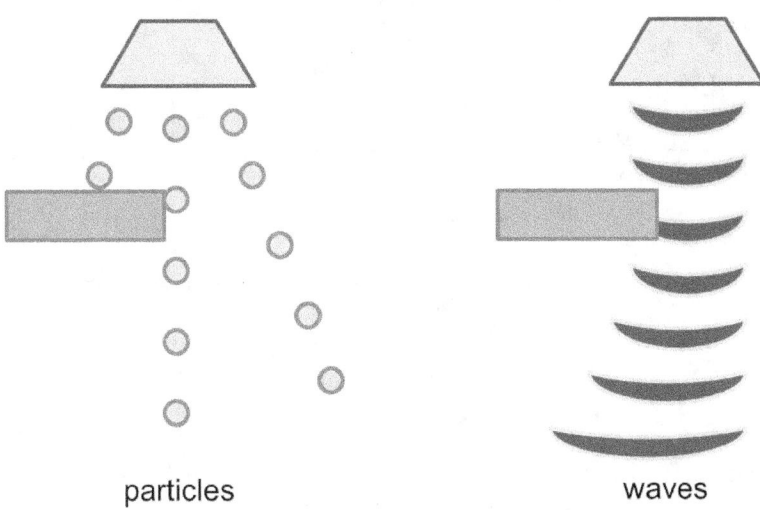

Figure 14- Diffraction of particles compared to waves

Figure 14 shows border diffraction, or not-diffraction, depending on whether photons, or electrons, are portrayed either purely as a particle or a wave.

The next figure shows the behaviour of photons and phonons in their diffractive effect after they have passed through a gap. With sound waves, it is the air that experiences pressure changes as the sound wave moves through it, and the pressure changes of the waves that move them around corners due to the drop in pressure as the waves "peek" around corners. In the following image, we see that higher frequency sounds bend less. Just as we also see in the adjacent accompanying image that higher frequency photons also bend less. Same behaviour for a similar reason. The medium supporting the waves is distorting at these boundaries as the waves trigger local density/pressure changes.

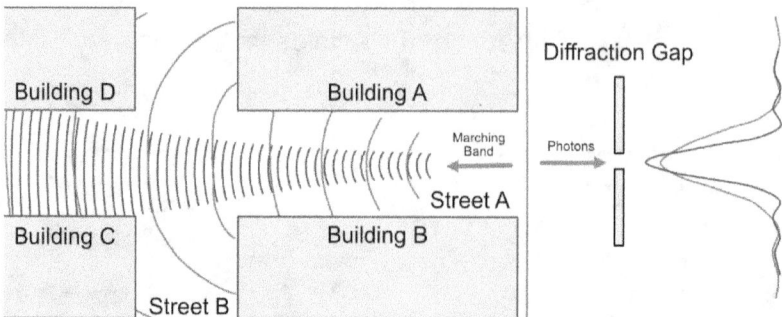

The longer wavelength sounds of musical instruments, of say a high school marching band, like the Tuba, diffract around corners more readily than the shorter wavelength, higher-pitched, sounds like those from another instrument say like the piccolo. Which presents itself by listeners coming down a perpendicular street hearing the lower frequency sounds more readily. Similarly, the longer wavelengths of red photons, diffract much more than the shorter wavelength blue photons where the gap separation is closer to the wavelength of the red photons.

Figure 15- Diffraction of low frequency vs high frequency sound waves and photons

Figure 15 on the left side shows what is experienced in terms of how different the frequency bending response is around objects for sound waves. In this case the edge of a building. Higher frequency sounds bend less. The right side of the image shows how blue light is bent less than is red light once it passes through a gap.

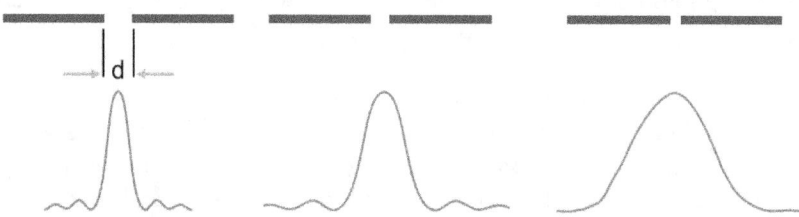

Figure 16- Single slit diffraction of photons

Figure 16 is a repeat of Figure 11 which shows that as the gap 'd' becomes smaller and thus approaching the wavelength of the photons being used, that the peaks and troughs of probability spread out. Here for easy reference for the reader of the e-book or paper book.

Apply this billowing out behaviour of the aether making up the pilot waves traveling ahead of the photons, or electrons, to the two-slit experiments, and we now have a mechanism for the so-called mysterious patterns observed. First pointed out by Louis de Broglie, at the beginning of the last century, and then later by Sid Deutsch. Not mysterious at all and a simple explanation as they follow the densest path, best conductor, ahead of them. Occam's razor should apply.

Phonons and photons show remarkably similar behaviour under similar circumstances. Including their speed being mediated by a conductive medium, whereas without a conductive medium as an explanation - Einstein concluded that "we simply have to accept this fact despite the failure of Michelson & Morley's to detect an aether, and move on." [I'm paraphrasing Einstein on this point.] "And accept that the speed of light is a constant."

An important excerpt from one of Lawrence Klaus's videos is where he was explaining to someone that "after removing everything from a volume of space it still somehow has mass." An obvious clue. But prejudice, and the flaw of the argument of authority prevents most from even considering a conductive aether. Yet almost everyone accepts the Higgs field interacting with matter via the Higgs boson and Higgs mechanism to give matter mass. How much more complicated does it have to get? It's already ludicrously complex. Ironically the Higgs field, Higgs mechanism and Higgs particles accomplish at a higher degree of complexity than what can be achieved with a universe propagating compressional waves where protons are standing-waves[2] in a compressible aether. [See the next figure.] The Higgs field and mechanism have become a moot point since the effect is a natural part of longitudinal density-wave behavior, and the Higgs particle is not necessary. The discovery of the Higgs particle is at best the mistaken identity of a "temporary high-density node" particle that is around the right corrected-guestimated mass, add to that they only exist for a tiny fraction of a second, and thus they are not the entities they claim to have discovered. See the book – The Higgs Fake (Unzicker 2013) by Alexander Unzicker.

The Higgs model of how mass is achieved actually is very similar in nature to the FOS gradient model. In the following figures, we see how the Higgs field and Higgs particles accumulate around so-called standard particles just as compression-density gradients achieve something similar and at far less complexity. There is no need to invoke the existence of additional particles, fields and a new mechanism.

---

[2] Note that I am not the person who came up with the idea of protons as some sort of standing-wave. I ran across this in some document about 30 years ago. I cannot currently find the original source to quote them.

The Death of the Dark Energy Idea

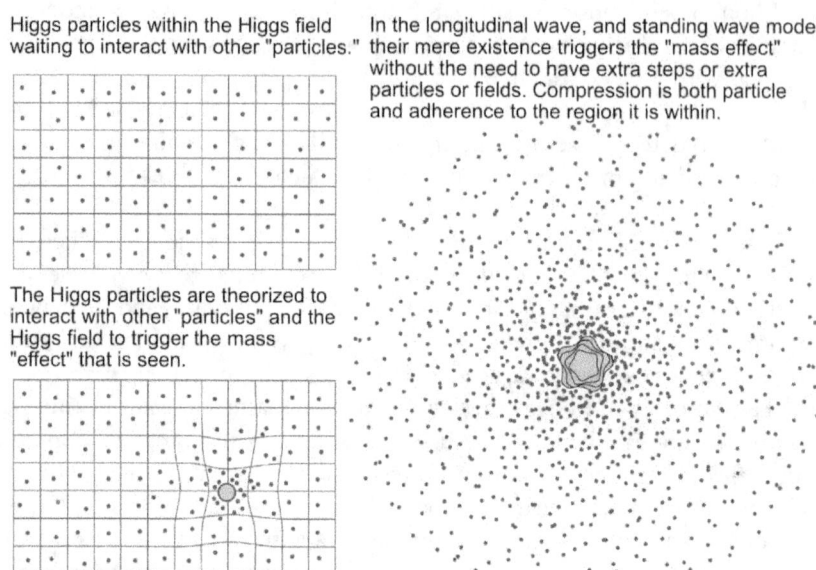

Figure 17- Higgs mechanism versus mass by standing waves

Figure 17 compares the complexity of the Higgs mechanism using the Higgs field interacting with matter via the Higgs particles. This field in theory is supposed to fill all of space with an enormous number of Higgs particles compared to the number of so-called real particles and through the Higgs mechanism mass-effect is achieved. While if there is an aether it simply fills all of space. And longitudinal compression effects give us a mass-effect with a much simpler and natural model.

The Higgs Boson and the "cocktail party" analogy

Figure 18- Higgs boson - the "cocktail party" analogy

Figure 18 refers to the so-called Higgs "cocktail party analogy." Where the crowding around normal particles by Higgs bosons within the Higgs field trigger mass-effects by the Higgs mechanism.

The Higgs boson, if found, is supposed to complete the Standard Model of physics. Its proponents speculate that matter obtains mass by it interacting with the Higgs field through its interaction with the Higgs boson. Via the Higgs mechanism. If the Higgs field, particle and mechanism did not exist, according to this model, then everything in the universe would be massless.

The cocktail party analogy is described as an event where until a regular particle shows up that the Higgs bosons are more or less randomly distributed around the room, with no one being the center of attention. The room is said to represent the Higgs field which is everywhere, and it is filled with Higgs bosons randomly distributed and moving about. Now a normal particle shows up, the celebrity guest, and the Higgs bosons are attracted to it like people are attracted to celebrities. As the celebrity wanders around the room, the other guests surround the celebrity, and this gives the group additional momentum. This mass of people has greater momentum and thus according to this model has acquired mass.

So, a "cocktail party" is used instead of a standing wave, or compression wave when in motion, for inducing a mass effect. Perhaps they are afraid of using the concept of a standing wave as it would imply an aether.

Michelson & Morley in 1887 failed to find the aether they were looking for because they were erroneously basing their idea on c+v where v was the extra component of velocity that the motion of the Earth added to or subtracted from its relative velocity with respect to the aether. How could they have not realized that an aether model still requires that the speed of light is still just c ~ 300,000 km/s. Whatever the medium is capable of. No more, no less. Just as sound waves are limited to the gas conditions, they are within, and no speed in the change of sound occurs for a moving train. Why would it be valid in the latter case but not the former for the speed of light? Alternatively, some people expected there to be forced flow of the fabric of space around an Earth composed of solid atoms. It was not until 1911 that Rutherford [et al.] discovered that atoms were not solid objects and appeared, in fact, to be composed mostly of empty space. Relative differences in speeds of stellar bodies moving through space are detected by redshift and blueshift changes, exactly like the Doppler Effect for sound waves, and has become an important tool in astronomy for studying the relative motion of the stars around us.

Einstein declared –"Just accept the fact that the speed of light is a constant." This constant velocity is also evidence for an aether, because any waves would be limited to the physics that such a medium would enforce, and everything else could not exceed the velocity that such a conductor can support.[Even Einstein's relative mass increase shows this on any particle that is accelerated closer to the speed of light as any such motion is unnatural and resistive to any velocity past normal conduction. There is no motion of c+v or c-v because you can only trigger variations in density during photon formation showing up as either a redshift or blueshift depending on the relative motion of the observer. No one expects to ride on the front of a train and start yelling and expect to find that the speed of sound changes for the yells. The medium conducts sound at only one velocity, and this would not be any different for the Fabric of Space – an aether.

Along with relative redshifts due to motion, an aether is also a mechanical explanation for a relativistic increase in mass that becomes apparent as the velocity approaches the speed of light. The FOS cannot support velocities of electrons, and non-photons, approaching the speed of light, so it begins to resist further compression and thus mass appears to increase. Einstein's relativistic mass equations support this new model.

Not only would an aether limit the speed of light, but the absence of an aether would imply that faster than light travel should be possible, and that we should be able to see photons traveling faster than the speed of light. The collision of photons with electrons never changes the speed of the photons only their wavelength and thus only the density/frequency changes, and thus again we have support for an aether.

AN AETHER / FABRIC OF SPACE VERSUS A MULTITUDE OF FIELDS AND PARTICLES

Given the current state of cosmology, it's easy to lose track of what today's "standard" picture actually requires. The Standard Model already treats space as filled with many independent fields, and modern cosmology layers on additional fields or substances to preserve the Big Bang framework (inflation, dark matter, dark energy, and related fixes). Whether one accepts those additions or not, the key point is that the dominant models do not describe empty space — they describe a vacuum populated by overlapping, space-filling fields.

The table below summarizes what is typically implied, and why that contrasts sharply with a single compressible, malleable medium model such as an aether/FOS (Fabric of Space).

	Field	Claimed to Fill All Space?	Experimental Status
1.	Higgs Field	Yes	Indirect (one excitation detected)
2.	EM Field	Yes	Well tested
3.	Gravitational Field	Yes	Indirect (metric geometry)
4.	Quantum Vacuum Field	Yes	Casimir is indirect evidence
5.	Gluon / QCD Field	Yes	Indirect only
6.	Weak Fields	Yes	Indirect via W/Z bosons
7.	Fermion Fields	Yes	Mathematical constructs
8.	Inflaton Field	Yes	Purely hypothetical
9.	Dark Energy Field	Yes	Inferred from expansion models
10.	Dark Matter Field	Yes (in field versions)	Inferred only
11.	Axion Field	Yes	Hypothetical
12.	GR Tensor Fields	Yes	Theoretical structure
13.	Unified / String Fields	Yes	Hypothetical

Table 2 A list of the major fields in modern physics

Below is a grouped and structured list, with expanded notes on what each field is said to represent in the Standard Model, quantum field theory, cosmology, or general relativity.

1. The Higgs Field
• Supposed to permeate all space uniformly, even in regions completely devoid of matter.
• Has a non-zero vacuum expectation value (VEV), meaning space itself is never in a "zero state."
• Responsible for giving mass to quarks, electrons, and the W/Z bosons through interaction strength (Yukawa coupling).
• Verified only indirectly through one excitation mode detected at CERN in 2012 (the Higgs boson).
• Mathematically behaves like a scalar field filling all of space with a constant background density.
• Functionally acts like a dense, elastic, space-filling medium whose resistance to excitation manifests as mass.

In effect, empty space is assumed to have a built-in mass-generating property.

2. The Electromagnetic Field
• Described as a gauge field that extends infinitely throughout the universe.
• Has zero rest mass, infinite range, and supports wave propagation at speed c.
• Exists even in perfect vacuum; vacuum fluctuations of the EM field produce measurable effects (Casimir effect, Lamb shift).

• Carries energy and momentum and has well-defined stress-energy components.
• Never truly "turns off" — even regions without photons contain zero-point electromagnetic oscillations.

In QFT, space is treated as an infinite electromagnetic oscillator system.

3. The Gravitational Field (General Relativity's Metric Field)
• Described not as a force but as curvature or tension in spacetime geometry.
• Exists everywhere — even in regions far removed from matter.
• Gravitational waves propagate through spacetime as real oscillations.
• The metric field stores energy and influences clocks, rulers, and light paths.
• Spacetime behaves mathematically like a deformable continuum with elastic properties.

Despite language differences, this is treated as a dynamic medium capable of bending, rippling, and storing energy.

4. The Quantum Vacuum Field / Zero-Point Energy Field
Includes:
• Vacuum fluctuations
• Virtual particle-antiparticle pairs
• Casimir-effect background energy
• Even "empty" space contains non-zero energy density.
• Virtual excitations constantly emerge and annihilate.
• Predicted vacuum energy density is catastrophically large unless artificially renormalized.

Conceptually this is indistinguishable from a structured background medium possessing density and fluctuation behavior.

5. The Gluon Field (QCD, Color Field)
• Represents the strong interaction between quarks.
• Exists everywhere, even in regions with no quarks present.
• Color confinement requires a non-trivial vacuum structure.
• The QCD vacuum is predicted to have condensates and non-zero structure.
• Infinite vacuum energy is predicted unless "subtracted out."

This is effectively a background medium with internal tension properties.

6. The Weak Nuclear Fields
• W boson field
• Z boson field

- Mediators of the weak interaction.
- Acquire mass through Higgs interaction but still permeate all of space.
- Responsible for beta decay and neutrino interactions.
- Present as continuous quantum fields even when no weak processes occur.

Even short-range forces are modeled as universal fields.

7. The Fermion Fields (Matter Fields)

According to quantum field theory, every particle corresponds to a field filling all of space:
- Electron field
- Positron field
- Muon field
- Tau field
- Three neutrino fields
- Six quark fields
- Even when no particles are present, the fields exist as ground-state oscillators.
- A "particle" is simply a localized excitation of its field.
- The fields never disappear; they merely fluctuate.

This means space contains at least 12 distinct matter fields at every point.

8. The Inflaton Field (Cosmology)
- Hypothetical scalar field proposed to drive early cosmic inflation.
- Supposed to fill all of space uniformly in the early universe.
- Has never been directly detected.
- Introduced to solve horizon and flatness problems.
- Acts as a temporary vacuum-energy source.

Another universal scalar medium proposed to exist everywhere.

9. The Dark Energy Field (Quintessence or Cosmological Constant)
- Proposed to uniformly fill all space.
- Responsible for accelerating cosmic expansion.
- Energy density remains constant as space expands.
- Mathematically indistinguishable from vacuum energy.

This is explicitly described as a cosmic background field.

10. The Dark Matter Field (Certain Models)

In some models dark matter is treated as:
- A scalar field
- A Bose–Einstein condensate

- A superfluid
- A galactic halo-filling continuous medium

These versions explicitly treat dark matter as a continuous, space-filling substance.

11. Axion Field
- Hypothetical spin-0 scalar field.
- Proposed solution to the strong CP problem.
- Expected to form a classical oscillating background field.
- Possibly constitutes dark matter.

Again, a continuous universal background.

12. Tensor Fields for Spacetime (GR)
General relativity models spacetime as:
- A continuous elastic manifold
- Capable of curvature and stress
- Able to ripple (gravitational waves)
- Possessing geometric structure at every point

Functionally this behaves like an elastic medium.

13. Unified and Extended Fields (Still Proposed)
- Einstein–Cartan torsion fields
- Scalar–tensor gravity fields
- Dilaton* fields
- Supersymmetric partner fields
- Superstring background fields
- M-theory brane fields

Each of the above concepts proposes an extended, continuous, space-filling entity with geometric or dynamic structure.

*A dilaton field is a hypothetical scalar (spin-0) field that appears mainly in string theory and some alternative gravity models. It is assumed to fill all of space and to control the strength of interactions — meaning that what we call "constants" of nature (like coupling strengths or even the gravitational constant in some models) could actually depend on the background value of this field. In string theory, the dilaton sets how strongly strings interact; in scalar–tensor gravity theories, it can determine the effective strength of gravity itself. In some extra-dimensional models, it represents fluctuations in the size of hidden dimensions. Experimentally, no dilaton has been detected. Searches look for deviations from General Relativity or variations in fundamental constants, but so far there is no confirmed evidence. Conceptually, however, the dilaton is another example of a continuous, space-filling field introduced to make high-energy theoretical frameworks mathematically consistent — effectively adding yet another structured background medium permeating the universe.

The Standard Model is often described as a theory of about 25 "fundamental particles," but that language hides something deeper. In quantum field theory, particles are not primary objects — they are excitations of continuous fields that fill all of space. Every quark, lepton, gauge boson, and the Higgs corresponds to its own space-filling field. When you count actual independent field components (including gluon color structure, electroweak components, spinor degrees of freedom, and the gravitational metric field), you quickly move from "~25 particles" to 60–100+ independent field degrees of freedom permeating all of space at every point. Add inflation fields, dark energy fields, axion fields, and various unification proposals, and the number grows further.

So while the public hears about a "particle zoo," the deeper ontology of modern physics is actually a densely layered vacuum of overlapping, continuous, interacting fields — effectively a structured medium filling the universe. By rejecting the historical concept of an aether, modern theory has not eliminated a medium; it has replaced it with a multiplicity of mathematically distinct but physically space-filling fields. From an Occam's Razor perspective, one coherent dynamic medium may be conceptually simpler than dozens of independent, overlapping space-pervading fields.

In regards to the Big Bang, what about the proof of the wall of the Cosmic Microwave Background Radiation [CMBR] that surrounds us? The currently most accepted account of the formation of the CMBR comes from about some 380,000 years* after the big bang. Over 13 billion years ago. As part of what is called the recombination event, or epoch, where the universe finally became cool enough for electrons to unite with protons to finally form stable atoms, and in the process released infrared and "deep red" photons. Since that time, they have shifted/expanded by a factor of about 1100. The temperature was hypothesized to have been about 3000 degrees Kelvin to just allow atomic formation, and since then has cooled to what is now the observed around 2.7260 (+/-0.0013) K, and will, in theory, continue to cool as the universe expands. The main problem with the CMBR origin model is that no matter when the Big Bang theorists say it came into existence, and what was the diameter of the universe when it was generated, is its continued persistence as this is a dispersion problem. The CMBR persistence problem. They have also had to modify every prediction about its intensity and uniformity, in order to have some hope of simulations generating the formation of galaxies in the right amount of time. Now they have given up on their original model, and are invoking dark matter and black holes to fix the problem.

The Death of the Dark Energy Idea

Once again if the model were to be true, then it is providing us with some odd position in time, and space, that makes this time unique. It should not continue to exist. Microwaves are photons and like any photons travel, more or less, in straight lines. Run the big bang explosion as a simulation and these photons should have passed by us long ago. Their presence right here right now implies that the universe was the diameter that we perceive it to be shortly after the recombination event/epoch occurred. As this is the only time it could have happened and yet still allow us to detect those photons uniformly around us, without a visual distortion in what we see for the CMBR wall, and that those photons have traveled in straight lines to get here. That means the diameter of the universe when the energy for the Cosmic Microwave Background Radiation was released when it was already some 28 billion light-years across. Exactly how far back we can see now. So much for the radius being "small" when compared to when these photons were supposed to be released. Once again if today's most popular theories hold true, they put us in a unique point in time and space – making us the center of the universe – which wreaks of error in reasoning. The James Webb Space Telescope, seeing more of the infrared part of the spectrum, will simply let us see more of the actual sources of the photons believed to be from CMBR boundary, and in the minds of too many, solidify our special place in the universe even further.

In the next figure we see that the some of the photons, o_n, that start in the outer regions travel directly to the core, while those photons, a_n, starting in the core start immediately after formation to move outwards from the core.

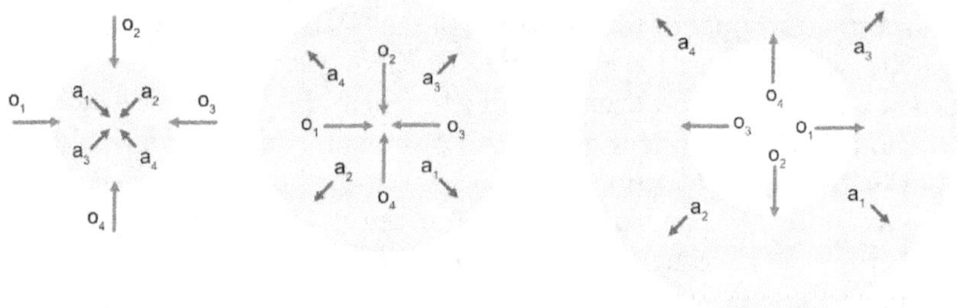

Figure 19- CMBR some 380,000 years after the Big Bang to CMBR void formation

Figure 19 shows the Cosmic Microwave Background Radiation starting roughly at some 380,000 years after the Big Bang and then its expansion overtime. Until finally at some unknown point in time a void showing no CMBR activity would have to form. But no such void exists yet. Nor any CMBR differential in its distribution. It's too evenly distributed unless we occupy some unique point in time and space.

The microwave photons currently making up the CMBR wall seem to be being continuously generated, or they come from so far away that we cannot determine their source. Eric J Lerner, and others, think it may be continuously being generated as a result of processes produced by our own galaxy. Perhaps from the magnetic field that is generated by and surrounds the Milky Way. My gut tells they are more likely from galaxies so distant that their related photons have become so diffuse and redshifted, that they cannot be combined and used to show us a common source. Here again, the James Webb Space Telescope will help provide a better map and idea of what can be detected. More distant galaxies whose photons have been extremely redshifted. Some are claiming that some of the hot spots in the WMAP are likely other universes pushing up against ours. A more straightforward explanation is this is simply microwaves coming from concentrations of stars that are too far away to be seen in the visible light part of the electromagnetic spectrum.

Another source of data that shows us that something is wrong with the reasoning of the CMBR comes from what is called the Hubble tension. That is the discrepancy of using the Cosmic Microwave Background Radiation data to derive the expansion rate of the universe (the Hubble constant, Hc or H0) versus using stellar Cepheid variable star (and Type Ia supernova) observational data from remote galaxies. This approach of comparing the two is often referred to as comparing early-universe methods (Planck satellite data for the CMBR) producing a value ~67 km/s/Mpc versus late-universe methods (direct observational stellar data) producing a value ~ 72-74 km/s/Mpc. A difference of about 5-6 km/s/Mpc which is considered quite significant. It is considered one of the biggest unsolved problems in cosmology. Not anymore! It is an unsolved problem if your entire premise of the CMBR is based on the recombination event 380,000 years after the supposed Big Bang. Why? It never happened. The perceived expansion, the implied velocities of ever more remote galaxies from their redshift, is simply an error in reasoning. It arises from using the wrong model to understand photons.

A switch to the model of a photon as a compressional wave gives us a much simpler solution in that they naturally spread out over time/distance unlike the transverse wave model of photons which requires the entire universe to expand to match the observed redshift captured by astronomers. With the naturally spreading out of longitudinal compressional waves, this means only the motion component can have additional effects due to changes in compression during formation and ejection/emission. The appropriate motion, compression or decompression, giving us the corresponding blueshift or redshift. Then there is No longer a need for space to expand, No mechanism for galaxies getting closer to the speed of light, and thus No need for dark energy. And with it - no Big Bang. While those who wish to keep things the way they are - are proposing additional particles and forces to resolve the most popular model's problems. The latter simply making it all the more complex and moving away from the principle of Occam's razor.

The accepted solution to Olber's paradox of why the night sky is not lit up like the surface of a star is the short lives of stars, and obviously the natural redshift of photons. This explains why our visual perception of the universe is limited. Referred to by some as the Horizon problem.

The gradient that is formed by a longitudinal/compression wave is identical in nature to the gradient of an atom's electric field in the Fabric of Space model. And the induced motion or displacement of the pre-compression and post-rarefaction pilot waves, or aether, of such a wave is by definition part of the magnetic field.

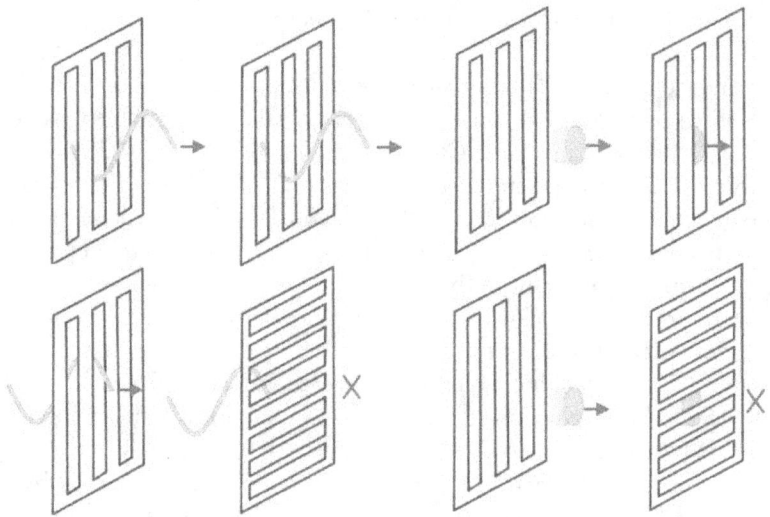

Figure 20 - transverse wave and longitudinal wave polarizations

Figure 20 shows the standard sine-wave string passing through two filters and being blocked by the second filter when it is turned at a right angle with respect to the first one. Similarly, it shows that if a photon has a elliptical profile that it too would have the same behavior, given that is also has a different width respect to its height.

Photons with elliptical profiles gain the ability to be polarized, and a comparison of the different wavelengths after their passage through polarizing filters is also proof of this. The shorter the wavelength the harder/less efficient it is to polarize photons, and thus the higher the frequency of the photons the more of them pass through. The polarization of longitudinal waves that have an elliptical profile better explains polarization then do transverse waves. The percentage of photons that make it through polarizing filters, is on the order of 1/3 of them, and this fact supports a height-width profile not suggested by a transverse wave model, a 3rd filter between two at 45 degrees to the others, who are 90 degrees apart, supports their fluidity of movement, and add to this them being preceded by de Broglie's [and Deutsch's] pilot waves give us a mechanical solution to the two slit quantum experiments. These key ideas, along with two-photon gamma-ray physics, and the clear evidence of the Compton Effect, for the malleability of x-rays, leads to an entirely different and simple model of nature. And the new electron model, introduced in the next chapter, alone provides a solution to the strong nuclear force, the weak nuclear force, electromagnetism, gravity and quantum mechanics.

Halton Arp's galactic-quasar triplets, a quasar in front of NGC 7319, NGC 7603 + 5 quasar bridge, are clear proof that we don't understand redshift completely, and the measurement of the twisting of space, by various groups, is evidence of the sheering of something. [Or perhaps the Earth's pilot wave having an influence.] Beginning first with Ignazio Ciufolini [et al.] using the data from the laser ranged satellites LAGEOS and LAGEOS-2, and later confirmed by Gravity Probe-B.

To ignore the physics required for the propagation of transverse waves because you have no explanation for it is more than just poor science. How good can your scientific method and thus your model be if you ignore data that is inconvenient? This has led us down the path of being unable to understand how the photons could be possibly so high redshifted and our conclusion based on this model is that the speeds must be getting greater and/or the universe is expanding. The alternative to this is simply that longitudinal waves naturally spread out over time and distance. Not because they are losing energy but because they are literally just spreading out. Of course, motion-induced redshift can still be trigged, but these two causes do not need a hypothetical dark energy. So, the whole dark energy requirement is totally bizarre in comparison to photons simply spreading out over time. Which is crazy! Then because of the problem of what is called fine-tuning the anthropomorphic principle is often invoked, or just as unlikely and distasteful is that quantum mechanics is invoked in a manner which they say triggers multiverse formation. With some science, and science fiction, writers saying that for every point in time where you had to make a choice that in fact both choices were made and two universes came into being because of that. While others have concluded that in most universes, the conditions are not conducive to the existence of life as we know it. And thus, they avoid the fine-tuning problem by saying we just happen to live in the right universe where the conditions are correct. For those who have elected to rejoice in such a formation of multiverses, what the have failed to realize or to tell you is that you would not likely exist at all. They have failed to consider that in such a succession of alternate events that due to the limitations of the biological lottery of life, that your very existence is unlikely. What is the average life span of a sperm and that of an egg? Sperm reportedly can live up to two to five days inside a woman where they are provided with nutrients from her to help them stay alive as long as possible. With between 30 million and 100 million per event. Where instead of you pursuing a career and becoming a famous "whatever", the reality is what they could say that might be true for your alternate "you" is also true for your parents, grandparents, etc. backwards in time. So, the likelihood of you existing at all in the multiverse becomes infinitely small in comparison to all the other possible egg-sperm combinations that would be more likely to arise if your parents chose to make love not just on another day, but even just an hour earlier or later. So, much for you having multiple lives in other alternate universes. Pure nonsense.

Even among those who support one another, with the most popular and widely accepted theories related to matter and the cosmos, there is growing dissent. Because the problems are becoming more apparent.

No black holes can form based on physics professor Laura Mersini-Houghton's work. "I'm still not over the shock," said Laura Mersini-Houghton. "We've been studying this problem for a more than 50 years and this solution gives us a lot to think about." [University of North Carolina News Archives]

Some plasma and electrical engineers supported her conclusion even before she had one. Like Hannes Alfvén, Tony Peratt, Eric J. Lerner, Donald Scott and others.

Anthony Peratt ran simulations of electromagnetism on galactic scales for the formation of galaxies and quasars and showed that gravity cannot explain the formations as easily, while simultaneously the Electromagnetic [EM] model can also produce the flat velocity profile of the motion of stars around their parent galaxy. There was no need to invoke dark matter. In 2019/2020 galaxies were observed that required no dark matter factor for their existence. Peratt's simulations also showed that no energy-and-matter ejecting supermassive black holes were required. The need for energy-and-matter ejecting black holes is in itself, for the most popular cosmological model, a contradiction of their definition. Ejection supposedly occurring as the matter is compressed just before disappearing into a black hole's event horizon. Conversely, both laboratory experiments, and simulations, show impressive results just using the dynamics of plasma. Including the prediction of pairs of "black-hole-like" currents at the cores of galaxies. Long before astronomers reported the apparent existence of pairs of black holes at the cores of several galaxies.

Without a big bang, many other theories are in trouble as well, such as the existence of magnetic monopoles, dark matter, and dark energy. There is simply no need of them anymore. Eric Lerner sums it up as; "If the Big Bang hypothesis is wrong, then the foundation of modern particle physics collapses, and entirely new approaches are required. Indeed, particle physics also suffers from an increasing contradiction between theory and experiment." From his book: The Big Bang Never Happened.

A list of some of the problems related to the Big Bang model: [Will go into greater detail at a later time.]
1. Horizon Problem.
2. Flatness Problem.
3. Matter cannot be created nor destroyed.
4. Galaxy formation, galaxy clusters show non-uniformity.
5. Inflation theory violates the speed of light.
6. Big Bang model now tied to dark matter and dark energy.
7. Galactic structures at the edge of visible universe.
8. Magnetic monopoles not found.
9. Others…

I am a reductionist to the extreme. And the new model I am proposing in this book does not support many of the desires of science fiction writers. But it does support most of the Electric Universe model. The EU model is gaining more support the more we detect, observe, and acknowledge the role of electric plasma currents and magnetic fields in our galaxy and elsewhere. Where once most scientists actually refuted the abundance of plasma and its importance in the universe it has now become so-called common knowledge. What is required is a merger of electricity and gravity, and now in this book, I will introduce a model that shows they are different aspects of the same phenomenon. They only vary in degree on an electron's orbital pattern with the gradients formed around nuclei, and the nuclei's position within the influence of the gradients of much larger masses, collection of nuclei, like that of the one generated by the Earth.

Chapter Two

Empirical Measurements

Any reasonable theory on the structure of the universe must be based upon the physics of nature that we can observe, directly or indirectly, which is based solidly upon empirical measurement and backed up by mathematics (orderliness). And although an equation is a testimony to our understanding of some aspect of physics, we must not forget that it is just a descriptive model. It in itself is not – does not create reality. Consider the successes of the ancient epicycle model of perfect circles for the orbits of the planets around the Sun, its ability to explain the apparent retrograde, the periodic appearance of the backwards motion of the other planets when viewed from Earth, and in the end its' complete and utter failure to represent the true motion of the elliptical orbits of the planets centered around the Sun and not the Earth. The epicycle model dominated astronomy for centuries. The failure to realize or apply limits to equations can lead to a complete failure in predicting the outcome of some scenario. For example, the standard equation of gravitational attraction fails when used to model the merging of two planetary masses. The general equation is only valid until the two planets make contact. A computer model that does not deal with this reality results in an incredibly inaccurate simulation. Another example is the equation for Ohm's Law. If you do not take into consideration the limits of validity then, for example, you might come to the conclusion that the electrical current would approach infinity for any applied voltage, along some electrical conductor, as the resistance drops towards zero. Instead, what happens is the system in question simply approaches the maximum current possible, that the given number of actual electrons available allows. Infinite current cannot be achieved when the number of electrons is limited in the system under consideration. Equations in engineering only work within the physical limits of the reality of the environment for which they were developed, and thus should be applied with this in mind. Any application of an equation, outside these limits is meaningless to the situation at hand, and terribly misleading to those who are unaware of the consequences of failing to work within the limits and thus produces useless results that are at best confusing.

This book is an exploration into a model/theory that attempts to describe the fundamental nature of the physics of the universe from simpler concepts; electro-magnetism, longitudinal compression-density waves and the physics of fluids and gases. The actual mathematics is not normally used to describe the model; instead, the basis of reasonably well-understood phenomenon are used as a means of applying equivalent phenomenon for the portion of the theory being discussed or described. As we discussed in the previous section, this work is based on an old concept of a luminiferous aether. Michelson & Morley proved that the aether, and particles, as they hypothesized them to be in their day [1887] did not exist. This point requires repeating – they failed to find the data to backup their model and a dark-ages model of the structure of atoms. The Earth was not dragging it, and their experiment could not detect a stationary aether that the Earth pushed itself through or that flowed around the Earth as it rotates on its axis because they were treating photons as transverse wave particles and not being conducted by an aether. The key idea behind their experiments being something that the aether was something that the Earth pushed itself through and thus triggered a flow of the aether around the surface of the Earth. Many have pointed out that their aether was preposterous because an aether that conducted transverse waves would have to be impossibly stiff and yet allow particles that make up matter to pass through it freely. Contradictory ideas. They were looking for a change of around 30 km/s relative to the Sun. The rotation of the Earth at the equator is moving at some 1675 km/hour or 465 m/s. While the Sun itself is moving around the Milky Way galaxy at about 828,000 km/h or 230 km/s. Experiments at the end of the twentieth century detected something that appears to be the twisting of space around the Earth [aka frame dragging]. Experiments done by Ignazio Ciufolini [et al.], Gravity Probe B and possibly by Roland De Witte. The whole speed of light issue was supposed to have been put to rest with Einstein simply stating that the speed of light is a constant. End of story. Well, yes, of course, if there is a luminiferous aether pervading all of space then the speed of light would be a constant, and you could only trigger variations in its density thereby resulting in either a redshift or blueshift of any generated photons. Then what is the story regarding Michelson and Morley?

When Michelson and Morley first performed their experiments in 1887 our idea of atoms was that they were solid objects. An idea, starting with the Greeks Democritus and Leucippus, then further developed by John Dalton, who had decades earlier [1803] proposed his Atomic Theory of Matter where atoms cannot be subdivided, created or destroyed. Michelson & Morley were locked into the concept of the Earth as a solid body that was composed of solid atoms forcing its way through outer space. Without any thought of the fact that the Earth is within the gradient of the Sun, and that we move within it. It was not until 1911 when Ernest Rutherford [et al.] showed the world that atoms were mostly empty space. Regardless of the practical radius you choose, Bohr or empirical, for the atomic hydrogen volume, versus the size of a single proton, the average hydrogen atom is about 99.9999999999996% empty space! Basically nothing! If I told you I had removed 99.99 percent of something you would think I had nearly eliminated something. If I told you I had removed 99.9999999999996% of something – you could hardly make the case that anything was left! If we took a cubic kilometre, of any material, and reduced it by this amount we would end up with a volume of about the size of a cube of sugar for a cup of coffee. This shows us how small the nucleus is compared to the radius of the hydrogen atom. On top of that if space is nothing then not only why does it have measurable properties, but at the same time, how can Planck's constant imply that it has granularity. How can 'nothing' be incredibly fine-grained? Quantum mechanics which dominates physics in our technological world arises from this very nature of quantization from the granularity implied by Planck's constant.

Any aether based model is considered ludicrous at best, and yet we are being told that the brightest minds have accepted a Higgs field, which interacts with other Standard model particles to give them mass through the Higgs mechanism as they interact with the surrounding Higgs bosons or particles that would need to be the most common particle in the universe after dark matter. All of which is itself embedded within a dominant matrix of dark matter [23%] which is then embedded within an even more dominant matrix of dark energy [73%]! Really?! This accepted model of the stuff of space is so much more complex than an aether model that it is ludicrous. This level of complexity is not necessary as the new aether model within this book will show you and how simple it is and surprisingly able to cover most of physics including the four* fundamental forces. Todays accepted models makes any aether based model look simple in comparison to all the extra particles and mechanisms required for accepted models to work. The sad thing, or funny [depending on your perspective], is that there are critical empirical and astronomical observational data that is being ignored because it does not support the more popular models, and thus this data is literally shunned because of what it implies. That is that our most popular models cannot account for the apparent redshift of some stellar objects that appear to be much closer to us than expected. Such as low redshift galaxies that seem to be associated with pairs of higher redshift quasars. This data collected by Halton Arp, if it holds true, implies that redshift is not completely understood and that some key ideas regarding the expansion of the universe, based on these assumptions, are just plain wrong. There is a much simpler explanation for all of it. The constancy of the speed of light implies an aether as its conductive medium, the increase in the mass of matter as it approaches the speed of light suggests an aether due to the experienced inertial resistance to change in conduction and similarly the effect of mass as well. Planck's constant, and with-it quantum mechanics implies an aether, and you will also see that not only so do electron orbitals, the photoelectric effect, but also gravity itself implies an aether. But not Michelson & Morley's aether - where the Earth was supposed to be a solid object that pushed its way through space. Without any thought of the fact that the Earth is within the gradient of the Sun, and that we move within it, providing a mechanical solution to Newton's mysterious action-at-a-distance. Long ago, we gave up the idea of solid atoms. Now we accept the idea of matter waves, and with it that matter and energy are one. Instead of solids, we now look to force fields created by charges, and now we have a simple story where wave-triggered density gradients are the key and that different kinds of waves make up everything else.

The key piece of evidence for this new model comes down to the following – the phenomenon called Two-Photon physics. This is where two gamma-rays interact with one another and give rise to an electron and a positron. We will look at this in greater detail in chapter three. The resulting "collision" of the waves gives rise to two of the most fundamental particles of physics with charge and mass! Somehow from two photons without charge and mass we end up with two particles that have charge and mass. How can that be? Are electrons and positrons particles? One of the key questions of science has been – is an electron a particle or a wave? The answer is it depends on how you define it. The distinction between the two is a perceptual illusion. We can design an experiment to choose to observe particle-like behavior or wave behavior. Whatever we want. When we want to see waves, then the experiment is designed to see wave behavior, and when we want to see particles, the experiments are designed to react to the density of a wave. Arising from the nature of longitudinal compression waves and simultaneously showing the fallacy of modeling the fundamental nature of the universe through transverse ripple waves. Longitudinal density waves with an elliptical profile are polarizable, and provide a physical mechanism and explanation for the wave-particle duality. This wave nature needs to and can be applied to the positron as well. When an electron meets a positron, they revert back into two gamma-rays! What happens when you break a proton? You end up with the following decay sequence, with an intermediate step of a positive pion decaying into a positive muon, and then finally into a positron and more gamma-rays! The mass of the proton is somehow taken away by the emission of gamma-rays, but the positive charge remains with the newly formed positron. If we consider only the Standard Model of physics, then somehow one of the possible nine versions, based on quantum chromodynamics, of a proton arises. Yes there are supposedly nine different protons. That is something rarely spoken of. This is based on a proton having to consist of three fundamental particles, consisting of two up quarks and one down quark. Yet in a decay event of a proton beginning with the shedding gamma-rays, mass, we end up back to a single fundamental particle the positron. Of which there is only one. Regardless of which of the kind of proton decayed. The math does not add up! The antimatter-matter annihilation experiments by scientists, like Hans Mes and Jacques Herbert, show over and over that protons and neutrons, and their antiparticles, in the end get converted into pions. Which ultimately decay into electrons, positrons and gamma-rays. Charged pions normally decay first into muons, and muon neutrinos, while neutral pions decay into gamma rays. Quark conservation fails, and this contradicts the Standard Model not only in terms of disproving the fundamental nature of quarks by the failure of their conservation, but in questioning their very existence. Richard Feynman disputed Murray Gellman's interpretation of the data on the existence of his quarks. With an added level of complexity by quantum chromodynamics of not only requiring nine versions of protons, but also nine versions of neutrons. Remember, or if your new to this, that their property of *color confinement* negates the ability of quarks ever to be observed. A great excuse to avoid the need for any evidence for their existence.

"Once the problem is eliminated by an excuse, there is no need to reflect upon it anymore." – Erwin Schrödinger

In regards to the physics behind electromagnetic waves, as transverse waves, no restoring force to bring such waves back to the equilibrium position is ever mentioned. Yet, it is required for a transverse wave to propagate. Which also requires a layered differential to exist for an equilibrium boundary to restore the wave disturbed 'space' back to how it was before the wave passed through it. They simply choose to ignore it. To ignore the requirement of this force for transverse wave propagation is poor science. This bears repeating. Because no solution was found, the worst possible thing occurred by those looking at the problem; they simply decided to ignore it and no longer discuss it. Wow! The consequence of this has led to much of the confusion in cosmology and particle physics that we see today. How good can your scientific method, and thus your model, be if you choose to ignore inconvenient data? That is not a good sign.

The majority of cosmology today is based upon ignoring the required physics for the propagation of transverse waves. And then apply their only known means of decay, expand the entire universe, to get the redshift of transverse waves they need for their model, and thus the need for the existence of dark energy to expand the universe.

The model presented to the reader in this book is no mere mental exercise in futility of trying to justify an aether. Because this mechanical model/theory does what the other theories have not been able to do – cannot do – it unites not only the Strong and Electroweak forces but also with them Gravity and Quantum Mechanics.

Michelson & Morley, via their interferometer experiments, did indeed prove that the universe they were speculating about was not the universe we inhabit. But that is the key point; they proved that their version of an aether universe was not supported by their experiments. Or there was a flaw in their logic. The twisting of space, around the Earth, has been verified by a few experiments. Performed by Ignazio Ciufolini [1997-1998, et al., Gravity Probe B [2004-2006] and Roland De Witte [1991] [aka frame dragging]. With some speculating that this shows some level of entrainment, but more likely density gradient sheering. In the FOS, fabric of space, model what is important is not entrainment, but density. What we are likely detecting is either the sheering of the density gradient formed around the Earth, or the changing of the gradient of a region as the Earth passes through it. That is, is it the Earth's pilot wave triggering a change? The density gradient is the gravitational inducing field. The same gradient that Newton was pondering and that gives the appearance of spontaneous action-at-a-distance. And the same gradient developed for and implied by Einstein's curvature of space equations. Remember they are still based upon the gravitational constant, which is the factor that is based upon the gravitational attraction of any number of bodies that are charge-neutral masses. Space-time is what we talk about, but equivalently we could talk in many of those cases simply as Space-density. Where the density changes the resonant frequency of atomic clocks. See the following figure comparing the curvature of space, and a density profile.

Einstein's general theory of relativity implies that gravity is a distortion of space-time and is graphically represented as a gravity well. See the next figure. This distortion in the space-time fabric is generated by the presence of the mass. This degree of complexity for an explanation is not necessary. Instead, we have a density gradient node in space, triggered by the mass at its center, but that is all. The time effect is simply due to the fact that the resonant frequency of an atomic clock is affected by the density of the material around it, or generated by its motion through a resistive Fabric of Space when in motion. Both effects predicted by Einstein but they are equally derivable from a solution based on density. There will be no tears in the space-time continuum as desired by some science fiction writers. Add to this the incompatibility of his curved space-time with the Higgs mechanism acting through the Higgs boson and the Higgs field to give us mass – and they have more than one problem.

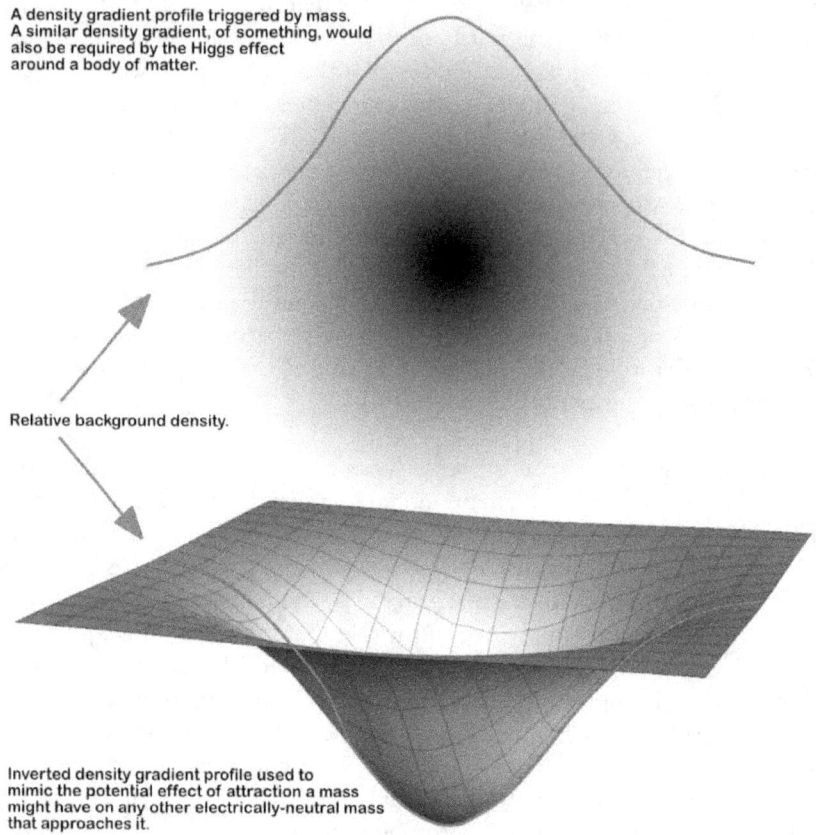

Figure 21 - Density gradient profile equivalent to distortion in space-time

The constancy of the speed of light is a red flag that there is a medium responsible for conducting photons, as waves in a medium are limited in velocity by their conduction mechanism. Inertial mass can be explained as standing-waves trigger a gradient by compressing it around themselves, or create a stable gradient shape due to uniform motion in such a medium. Changes are resisted and require an active force to change the gradient form. The Higgs field and mechanism are almost identical to a solution using a compressible aether, but far more complex in requiring a Higgs boson to interact with matter to produce mass. Adding that atoms see an increased resistance in gaining additional velocity further supports such an aether and compliments Einstein's relativistic increase in mass at the same time. The original theory or view for a medium is that it would have to be incredibly rigid for the propagation of transverse waves at the speed of light and yet incredibly tenuous to allow the particles that make up, what was once considered a solid-atomic Earth, to flow through the aether without experiencing drag. These properties of rigidity and tenuous at the same time are contradictory in their nature. Let alone their propagating in straight lines in three-dimensional space without some two-dimensional like boundary to travel along. There is an answer.

How can an aether theory explain the issue of enormous bodies like the Earth, composed of a huge number of particles of matter, being dragged or forcing its way through an aether and yet show no signs of resistance? Remember they originally assumed that the Earth was a solid object and only later did they discover that in fact atoms are on the order of 99.9999999999996% empty space. Now let's go one step further - because there are No solid particles. There are only waves, photons - electromagnetic waves, electrons [toroidal "smoke ring" shaped waves], standing waves [protons and positrons], and variations in the density of the FOS. Their particle properties arise from the forces and fields they generate and how they interact with one another, how an experiment is setup to detect properties, and where localized high-density regions can act like solids and produce quantum effects. Where quantum effects are either quantization or statistical probability regions or events. Just as we know atoms are mostly empty space, and the solidity of masses of atoms is due to the repulsive action of the electrons between them. So, the aether does not have to be some impossibly dense solid that conducts transverse waves, and yet also somehow allows particles to pass through it freely. The particles, as currently envisioned in the most popular models, don't exist and transverse waves are not the only polarizable waves. The majority of particles ultimately decay into electrons, positrons, neutrinos, and various forms of EM radiation. These are the only stable forms, and anything that is unstable is Not fundamental. Fundamental implies stability. This creates a problem for the Standard Model once you realize it is not a Fundamental Model. It is a model based on the observed possible energies of particles, and their apparent masses while they exist outside of nuclei for tiny fractions of seconds, triggered in extreme energetic collisions between normal atomic matter. Not the conditions in which atoms normally exist unless they are in a star. On top of that today's most accepted model is rarely shown in full with all three types of each quark. There are three up quarks, three down quarks,… and this does not even take into account the supersymmetry counterparts of all the particles that the LHC [Large Hadron Collider] has failed to find. This has many theorists worried or saying we need a more expensive collider to find them. Ouch.

The reason longitudinal-compression waves were dismissed to model the structure of photons, and instead transverse waves were exclusively selected, was due to the ease of explaining the polarizable nature of light as transverse waves, and their apparent bending around objects when modeled as transverse waves on water. It's easy to visualize them as displaying their polarizable nature as transverse waves because of their similarity to strings vibrating up, and down, and passing through polarizing filters aligned with this movement, and blocked when the filters were rotated by ninety degrees. Experimental data does not support photons being simply two-dimensional waves. An elliptical profile may well be supported by the general nature of polarizing filters to eliminate 2/3s of the photons on average. Thus, suggesting that they are almost as wide as they are tall in a profile passing through the filter. Whereas if they were more purely almost 2D in their profile as is often suggested, then less photons should be passing through, and add to that demonstrations of their rotation through 45-degree middle filters would be even more problematic for transverse waves but not for longitudinal waves. Even more important is how does the current most popular model of physics move from two dimensional waves turning into an electron/positron pair as happens in the outcome of the two gamma-ray photons-colliding phenomenon. Two gamma-ray physics. It does not.

The failure of the theorists was to not consider longitudinal waves with an elliptical profile. Which in itself better explains the apparent twisting of photons through polarizing filters, not just quarter-wave retarders, as these photonic masses are guided through such filters. They are guided through as if their bodies can trigger their passage around the filter sides due to more of their mass making it by, then being guided by what appears to be a lower, barely detectable, density wave front of material forming ahead of them - pilot waves. The existence of pilot waves has been showing up over, and over again in various natural phenomenon and experiments. The idea of pilot waves is supported directly by Yves Couder's [et al.] silicone oil droplet experiments, from the University of Paris, showing quantum-like interference for two-slit experiments arising from the pilot waves that are generated in the fluids in the presence of the silicone oil droplets. These pilot waves providing additional properties that answer some of the questions we have regarding the various phenomenon. It was Louis de Broglie [1892-1987] who brought up the idea of pilot waves as a mechanism to account for some of the unusual effects seen in quantum mechanics. And they also have implications for John Steward Bell's theorem and Alain Aspect's accompanying experiments regarding non-locality and the appearance of faster than light phenomenon. Sid Deutsch points out a flaw in Aspect's experiments regarding their inability to take into account the polarization of photons and the effects the pilot waves would have on the data collected.

With respect to the twisting of photons through polarizing filters, this behavior seems reasonable for a compressional wave to move through polarizing filters not exactly lined up with them, no such simple explanation can be visualized for transverse waves which are always portrayed as being nearly two-dimensional single planar entities. And yet the properties observed for photons indicate behavior that is more indicative of longitudinal compression waves, not transverse boundary waves. Circular polarization, transfer of momentum, particle formation in two gamma-ray photon collision, photoelectric effect, electric field gradient, matter waves do not sound like transverse waves, spreading of waves after passing through a singular slit or two, the Compton Effect,... these all are more easily explained by longitudinal compression-density waves, in conjunction with pilot waves, not '2D' transverse boundary waves.

What triggered this latest exploration into a new model was partly due to the lack of physical solutions for some of the more basic questions about how the universe, or at least how some parts of it, appeared to work. In particular, while in high school, I wanted to know why the hydrogen atom was not neutral even though it was composed of one proton and one electron. Too many chemistry books implied that the reason that hydrogen atoms were chemically active, and not neutral, is that they "like to fill their electronic shells." Like?! "Like" is not an explanation! This does not make sense if you consider that you are uniting a positive and negative charge. They should have been neutral. You could point out that the neutron is neutral. But then you need to consider - Why is it then that outside the nucleus the neutron is unstable. Something else the quarks of the Standard Model cannot explain. The half-life of a neutron is around ten to fifteen minutes. Minutes! It's not a stable system. If anything, you would think that it would be, and the standard model does not account for it let alone predict it. Another motivation for this exploration was the ever-increasing complexity that is necessary to support, and explain, the most widely accepted theories of the structure of matter and the universe. Of course, I'm talking about String theories. Portrayed by many as an unquestionable fact. But string theory is not a fact. A few people have now written books bringing the problems with string theory to our attention. Like Jim Baggot, Lee Smolin and Peter Woit[3]. String theory is just a theory, now actually considered a group of theories that are attempting to come up with a model that can explain what we observe in the universe. And yet not one experiment exists that can support the existence of these string particles, or perhaps more importantly, no experiment can disprove their existence. Technically a theory has to be unprovable. Yes – falsifiable. Yet there is no way to disprove string theory. It makes no predictions. There is no empirical evidence for it. Nothing can be ascribed to string theory. String theory arose as an exploration into an attempt to explain the fundamental forces through the concept of tiny trans-dimensional one-dimensional strings, or particles, who we are told are not only incredibly tiny but exist in other dimensions but whose influence are felt in our world. Requiring, depending on who you talk to, between ten and twenty-six dimensions. By vibrating just the "right way" they are supposed to trigger the known forces. It is, or was, an attempt at a theory of everything. Of course, we don't know what that "right way," really means or how it is supposed to work to provide the forces. Not only is there no proof of what they say, but they are also unable to make any experimentally verifiable predictions that would support their work. A few authors have now written about their concern about the talent and money

---

[3] Jim Baggot – Farewell to Reality, Lee Smolin – The Trouble with Physics, Peter Woit – Not Even Wrong

wasted on string theories. Worse yet in accepting string theory's other dimensions has arisen in the minds of some in the general public the notion, and fanatical belief by too many, is that these extra dimensions is support for inter-dimensional beings, aliens inhabiting parallel universes, and other astounding ideas.

What proof or experiment could invalidate the FOS model? Karl Popper's requirement for the validity of a theory. One must be able to disprove it. Yes, find evidence against it. The most obvious ones are the containment of dark matter, dark energy, a singularity, a black-hole consuming one of the planets in our solar system, aliens pouring out of a wormhole or star-gate, faster-than-light travel, magnetic monopoles [not "something-like one" as has been reported by one group], time travel,... Experiments related to the FOS model could be used to investigate something more at the atomic level. Like the photon interference patterns for two-slit experiments affected by observation. The Observer Effect. Which appears to be a sensor's proximity to the gaps triggering interference. Likely preventing the normal aether distortion similar to wave-gap diffraction.

All the major/popular theories have become so complex. Long ago treading away from the tenet of Ockham's (Occam's) razor;- "that most terms, concepts and assumptions should not be multiplied beyond necessity.". In other words, when theories compete with one another, often the simpler of them has turned out to be a more accurate description of reality. A tenet that many hold to be a safe guide in reasoning.

Nobody can explain exactly how gravity works. Einstein's General Theory of Relativity introduces the notion that mass is a distortion in space-time and can tell us what the consequence of an accumulation of mass results in. But then the Standard Model is seeking out the graviton particle as well as evidence for the Higgs field and Higgs boson. Where mass is "passed on" to particles by their interaction with the Higgs field through Higgs bosons through the Higgs mechanism. Is the Higgs model an unquestioned fact? No. Alexander Unzicker is one of the few who tells us this is not so.

"A particularly worrying symptom of the current state of affairs in physics is the so-called discovery of the Higgs boson at CERN. The media-hyped announcement in 2012 has been followed up by a series of announcements, each installment making the case that the big sensation is "increasingly more likely." But what was actually discovered were a number of unexplained signals obtained by extensive filtering methods, raising many questions for everyone who takes a sober perspective. Nobody can claim to oversee the analysis of the massive pile of data produced by CERN's collider experiments. Nevertheless, these signals are pushed to serve as evidence for the long-theorized Higgs boson supporting the "standard model" of particle physics, although this standard model is not even a well-defined theory. Such an interpretation speaks more of desperation to validate the past six decades of research and to shore up a model that is wobbling precariously under the weight of all the bits and pieces glued onto it to make it work." [Unzicker, Alexander; Jones, Sheilla. Bankrupting Physics: How Today's Top Scientists are Gambling Away Their Credibility]

So, we do not have consensus. On top of that, all the formulae only tell us what gravity is capable of - what it can do. Not how it does it. From the existence or non-existence of gravitons, or gravity waves, their supposed discovery would still in no way explain how they work to unite bodies of matter. Another failure is the explanation for how the electrical force works supposedly by the exchange of photons between charged particles. The exchange of photons still does not explain how it unites or repels bodies. Even the most popular of theories are not as perfect as most of the literature suggests or adamantly state. Which is even more misleading for those of us who are not able to distinguish between concrete facts and just the currently most "popular and accepted" theories. Luckily for those of us who have not been turned off of, or intimidated by science, there are a few educated individuals that have taken the time, by way of writing books, and posting on the internet and other means, to explain to the rest of us about the realities, and their shortcomings, of some of those theories. People like Lee Smolin, Peter Woit, Jim Baggott, Alexander Unzicker, Halton Arp, Hannes Alfven, Anthony Peratt, Eric Lerner, Donald Scott, Sid Deutsch... to name a few.

Quark theory is one of those not so perfect theories that the average person with some interest in structural atomic physics has been led into believing that it is an unquestionable fact. At least this is how it is portrayed in most of the literature. But luckily not by everyone. The quark model fails in some of the most basic fundamental predictions that it should make regarding the nature of nuclei and atoms. The quark model does not predict the instability of neutrons outside the nucleus [a half-life of just over ten minutes], it does not predict the negative charge profile of neutrons within the nucleus, it does not predict any form of radioactivity, it does not predict the ability of hydrogen atoms to share and take on another electron, it does not predict the quantum/quantized nature of atoms, it does not predict the increasing wavelength of ultra-cold neutrons and their shorter lives, and does nothing at all in terms of gravitation.

It may have started as a theory to come up with a more fundamental nature of baryons [protons, neutrons] with the charge of them being obtained by proposing the existence of fractionally charged particles. The up quark with a charge of $+2/3$'s and the down quark with a charge of $-1/3$. Thus a proton is supposed to be composed of two up quarks and one down quark giving a total charge of $2x(+2/3) + (-1/3) = +1$. While the neutron is supposed to be composed of two down quarks and one up quark thus having a total charge of zero. $2x(-1/3) + (+2/3) = 0$.

At first, this seemed nice and neat until someone pointed out that this violated Pauli's exclusion principle, which says that no two identical particles can exist in identical states within an atom. The solution? Let's propose more types of quarks to resolve this. Thus arose quantum chromodynamics. They decided that there must be three sub-types for each main type. Thus arose three types of up and down quarks called red, green and blue. Odd names. But then when you are making up something, you can call it whatever you want. It's just a name for an idea. The problem is that this means there is nine types [versions] of protons and nine types of neutrons. Quite the complexity that is never really discussed. Nor does it cover what is called the Proton - Neutron distribution problem. That is where if you look at the distribution of the protons and neutrons within nuclei there is a problem of how close some of the protons get to each other and the repulsive forces that must exist in certain areas. Because of this the nucleus would have regions of high stress due to the repulsive positive charge between protons and is constantly in a state of being on the verge of tearing itself apart because of these regions.

The following tables show the nine different versions of protons and neutrons. Due to the theorized existence of the different types of quarks and quantum chromodynamics. We have Up quarks which in theory have a +2/3 positive charge [red, green, and blue], and Down quarks which need to have a -1/3 negative charge [red, green, and blue] in order for the theory to work.

Nine different protons formed from different quarks [2 ups, 1 down]

Up quark	Up quark	Down quark
Red	Green	Red
Red	Blue	Red
Green	Blue	Red
Red	Green	Green
Red	Blue	Green
Green	Blue	Green
Red	Green	Blue
Red	Blue	Blue
Green	Blue	Blue

Table 3- Nine different protons formed from different quarks [2 ups, 1 down]

The Death of the Dark Energy Idea

Nine different neutrons formed from different quarks [2 downs, 1 up]

Down quark	Down quark	Up quark
Red	Green	Red
Red	Blue	Red
Green	Blue	Red
Red	Green	Green
Red	Blue	Green
Green	Blue	Green
Red	Green	Blue
Red	Blue	Blue
Green	Blue	Blue

Table 4- Nine different neutrons formed from different quarks [2 downs, 1 up]

Quantum chromodynamics, the quantum dynamics of color interaction [not to be confused with color confinement], is used to explain how the Pauli Exclusion Principle is not violated and thus allow Up and Down quarks to explain how quarks are allowed to form protons and neutrons. [In Chapter 2 we look at low energy Electron Capture, and it's interesting to note that their theory must account for how an electron changes one of the proton's Up quarks into a Down quark to convert it into a neutron and it always correctly adjusts the color somehow. Unlikely. In fact, we have the added problem of somehow a fully charged electron converts, or changes, the quark into a fractionally charged body. Then disappears. In violation of a few principles.] So, remember when you are looking at the table of Standard Model of Elementary Particles that most of them do not show the three varieties of up and down quarks. The red, green and blue ones. Please don't confuse them with the three generations of quarks: 1st generation Up and Down, 2nd generation Charm and Strange, 3rd generation Top and Bottom. Try not to get hung up on all these names and varieties, as ultimately that which we can truly detect without bias are the electrons, positrons, neutrinos and various forms of radiation – the four end-types or truly fundamental entities [for a lack of a better term at this time]. Anything else either has not been actually seen or has existed for so short a period of time it makes one wonder how can it be considered fundamental if it lasts less than a microsecond and then decays into some combination of the main four end-types.

Quantum chromodynamics, not to be confused with quantum electrodynamics, is, in the end, the most accepted model that is supposed to provide us with an understanding of the strong nuclear force, and provides greater detail in how Hideki Yukawa's pion exchange model for the strong nuclear force is supposed to work at a more fundamental level of detail for greater clarity. It does not explain it any better, and in fact I would argue that it makes it even worse and adds a level of complexity moving away from the principle of simplicity of Occam's razor. Yukawa's model is complementary to the model in this book. As you may be aware another problem with the quark model is that they cannot be seen directly, although deep inelastic electron penetration of protons shows that they are not point-like objects, and to provide an explanation, or excuse, for why this is so was provided by the idea that the force between quarks gets stronger the farther apart they are while becoming weaker, the closer they are together. The force can become so strong that during any extraction process, or explosion, that a new quark-antiquark pair would be generated between the proton and the extracted quark triggering the reformation of the proton and the extracted quark would react with the antiquark to form a meson. Thus, keeping the mathematics of the model from being violated. Protected from violation by what is called color confinement*. Could it possibly be any more complicated! In the end, this provides an excuse as to why there is no need to show the existence of quarks to the rest of us. How convenient. As Erwin Schrödinger long ago said: "Once the problem is eliminated by an excuse, there is no need to reflect upon it anymore."

Richard Feynman was not convinced the data from the experiments supported Murray Gellman's quark model of tightly locked particles at the core of protons and neutrons, and instead described the data as suggesting loosely held partons much to the irritation of Murray Gellman. Fluctuating density is exactly what the FOS model theory states should be seen as the standing wave node [proton/positron] fluctuates, and rotates. And with that the appearance of locations within the node, or at the surface, of higher density regions coming into and going out of existence thus appearing as loosely held objects. [More on this in chapter 2.] The following image is not intended to represent the ideal standing wave model, but is simply meant to display that any particle/wave striking the outer region of a standing-wave like body sees more material at the outer perimeter and thus is deflected more in these regions than any particle/wave striking it head-on where it appears to have a lesser mass and thus passes through it more readily. [A proton also has a spin component generating its magnetic field.] Is this boundary the most significant region due just to the undulating body of the standing wave, or does some of the more highly compressed material at is surface play a significant role?

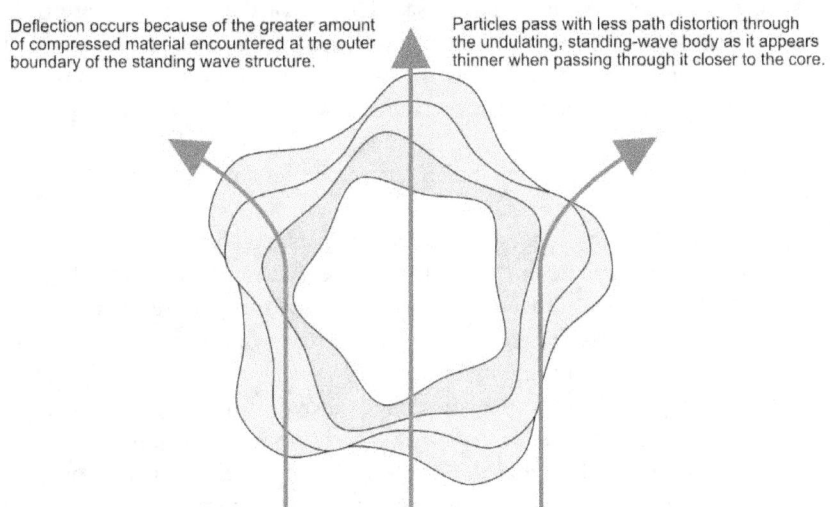

Figure 22- Standing wave with non-point like scattering

A standing wave body is seen thinner when particles or waves strike it head on versus hitting the more apparent thicker outer regions.

The evidence in support of the quark model used to be limited to the non-point like inelastic scattering experiments suggested for the nature of protons, but has now some additional support by the evidence of the existence of fractional charge of 1/3 discovered by an experiment using niobium balls. The only problem with this is that nearly every atom is fractionally charged to some degree. This is why chemistry works. We just discussed, two paragraphs ago, that the isolation of a lone quark is in complete opposition to this concept we were told about - that individual quarks could never be seen nor detected because of the property of color confinement. Originally the data showing that protons do not have a point-like nature was discovered at the Stanford Linear Accelerator Center in 1967, by Henry Kendall, Jerome Friedman, and Richard Taylor. They used the accelerator to bombard protons with electrons and observed what happened much like the Rutherford experiments that discovered the nuclear-core nature of atoms, and that they were mostly empty space. Based on the scattering of the electrons, they realized that protons were not point-like objects, and thus they were led to believe that they must contain smaller objects. Richard Feynman did not agree with Murray Gellman about what the experiments showed regarding the nature of protons and neutrons. "Still this was not taken as confirmation of the quark model. Richard Feynman, also at Caltech, analyzed the scattering results and concluded that whatever was inside the proton – he called the objects "partons" – seemed to be rattling around like marbles inside a tiny can. Gell-Mann's quarks, on the other hand, were supposed to be tightly, irretrievably bound, and from a strictly empirical point of view Feynman was unwilling to say that partons and quarks were the same thing. Gell-Mann tended to be disparaging about Feynman's partons, as if his colleague were merely trying to avoid using the word he found distasteful, but there were genuine grounds for puzzlement4."

A quote from The Atomic Scientists[5] enlightens us further to the reality of the state of quark theory:

"Despite the many successes' quark theorists have enjoyed in presenting an orderly and understandable picture of the hadron spectrum, there is no experimental evidence that the quarks are anything more than convenient mathematical fictions for expressing the known symmetries. We are certainly not justified at this point in concluding that a nucleon consists of three physical entities having the properties described above simply because the baryon spectrum can be explained in terms of such a model. In a sense, this would be equivalent to concluding that the photon is composed of an electron-positron pair because the unit spin and the zero charge of the photon can be derived from such a model. We are thus left with a tantalizing and incomplete picture in which the observations are arranged in a very orderly scheme based upon symmetry principles but give us no real understanding of what these symmetries mean for the dynamic structure of the baryon."

If they are having trouble with the building blocks, then no one should be that surprised by the lack of any clear understanding in how these particles interact with one another through the strong force, or the theory of this force that has been termed Quantum Chromodynamics (QCD). "With all the complex mathematical machinery and all the sophisticated physical assumptions that have been introduced to handle QCD and obtain deductions from it, the results have been very meager and at the cost of replacing simplicity by complexity. Taking color and gluons into account we see that QCD as it now stands is burdened with some 30-odd arbitrary parameters. We can hardly speak of this theory as a simplification or unification of physics. This complexity has been aggravated by the introduction of three additional quarks called charm, bottom and top, each in three colors, though these quarks play no essential role in the scheme of things."[6]

"QCD also leaves much to be desired in predicting the masses of the individual quarks and the magnetic moments of the baryons. It gives individual quark masses, which are nowhere near the empirical masses required to account for the measured masses of the nucleons. Nor can it give the magnetic moments of the proton and neutron. At best it gives the ratio of the two magnetic moments as 2/3 without accounting for the fact that the proton magnetic

---

[4] by David Lindley-from his book: The End of Physics
[5] by H.A. Boorse, L.Motz and J.H.Weaver
[6] from The Atomic Scientists: A Biographical History

moment is positive whereas that of the neutron is negative."[7] While in the FOS model we are presenting here, the neutron and electron are expected to have negative magnetic moments while the proton has a positive one. [More to come when the content can be paid for.]

In more recent times, 2013, Alexander Unzicker and Sheila Jones commented about the quark and color problem in their book…

"SIMPLER, BUT NOT REALLY SIMPLE As Murray Gell-Mann frankly admitted during a talk in Munich in 2008, Heisenberg considered the entire idea of fractional charges assigned to quarks to be nonsense. Had the genius Heisenberg already become a senile, stubborn skeptic at the age of 65? It is unlikely that he felt biases against fractional quantities, as he, in his freshman years, had proposed the famous "half-integer spin" on an electron, which back then stood in sharp contrast to the established wisdom. However, half-integer spins make sense observationally, whereas no one has ever seen a fraction of a charge.* This is perhaps the most absurd shortfall of the model: isolated quarks don't exist. Quarks always appear as pairs or triplets, and there is no way to break them apart into singles. No one knows why, although a neat term was invented to describe it: confinement. Does it make sense to talk about parts of something that cannot be divided into parts? Another very nasty problem arose because the quark model had to care about an exclusion principle in quantum theory found by Wolfgang Pauli. It forbids identical particles living together in the same place. To get around that problem, physicists had to invent another messy attribute in order to make it okay for identical quarks to live together in the same place. They invoked colors so that quarks living together would no longer be identical, but as Roger Penrose relentlessly reminds us, they are "in an essential way unobservable." The colors, in brief, were introduced solely to keep the theory from contradicting itself. And once again we face an example where contradictions were removed by inflating the model. In the canon of particle physics, mostly told in retrospect, each step seems reasonable, but it is sobering to follow the detailed account in the book Constructing Quarks: A Sociological History of Particle Physics by Andrew Pickering. Pickering is a highly qualified particle physicist, but his experiences in the field led him to investigate the mechanisms for how today's mainstream concepts were established. He demonstrates the way successive adaption of techniques that are sensitive to the desired effect greatly influences what is generally thought to be an experimental fact. The countless erroneous interpretations, contradictions, ad hoc assumptions, and widespread herd mentality described by Pickering raise serious questions about the entire method of high-energy

---

[7]from The Atomic Scientists: A Biographical History

physics.[8]" [Unzicker, Alexander; Jones, Sheilla. Bankrupting Physics: How Today's Top Scientists are Gambling Away Their Credibility]

Another interesting comment on the reality of the state of physics from the authors of The Atomic Scientists is that in "hadron structure we must not forget that what is perhaps the most fundamental and simplest question of all remains unanswered: What is the structure of the electron and why does it have its observed properties?"

It's the failure of all these mathematically complex theories inability to answer the above type of simple questions that motivated me into exploring a physical type solution for atomic physics. I wasn't really sure if anything could be gained by the exercise, but then again, I did not like what I was hearing. It was also the alienation of others as well from science that made the struggle a little easier. Their reaction to atomic physics reminds me of the old saying: "If you can't amaze them with your intellect, then flabbergast them with your bullshit." Something else that kept me going was some of the wild extrapolations some individuals were making from some fragments of various theories. Theory fragments like the "fact" that superstring theories require anywhere from 10 to 25 dimensions - "many folding in upon one another." With other statements that make you wonder if they are just grasping at any excuse not to require evidence. Like "dimensions so small" that they're not visible in our universe. What the hell? From Wikipedia: "In string theory, a model used in theoretical physics, a compact dimension is curled up in itself and very small (usually Planck length). Anything moving along this dimension's direction would return to its starting point almost instantaneously, and the fact that the dimension is smaller than the smallest particle means that it cannot be observed by conventional means. Extra dimensions in a theory which are made compact are said to have undergone compactification." Add to this complexity that there is no need for evidence, and experiments that have been interpreted to "indicate" variations and the possible slowing in the passage of time, and you begin to feel that something is not right. One piece of evidence for time issues comes from the observation that apparently muons are surviving for longer than expected as they enter the gravitational well of the Earth. Along with the variation in time based on changes in the oscillation atomic clocks triggered by motion, and being at various elevations, with respect to another clock for reference. From a few feet/metres apart to thousands of feet/metres above the surface of the Earth, or even below the surface of the Earth. A simple explanation exists for this as well. These observations have led some people into contemplating the existence of time travel, and various other fanciful speculations related to time. The main problem I have with popular theories like these is that they conflict with my own prejudicial three-dimensional world view. A view held, I believe, by the majority of Homo Sapiens Sapiens.

---

[8] From Bankrupting Physics: How Today's Top Scientists are Gambling Away Their Credibility by Alexander Unzicker, and Sheilla Jones.

Although Michelson & Morley proved that the aether they envisioned, with a solid Earth traveling through it, did not exist my own exploration did begin with a look at their experiments - to understand their view of an aether universe. I craved a reasonable explanation, and I had already conceived of two concepts that inspired me to explore an aether universe. To even hope of finding the solution to a problem where so many others more educated than me had failed, or given up, seemed at first preposterous. Let alone getting such a theory reviewed simply because of the ingrained prejudice against an aether model. And yet today we are told of an "aether" that is accepted by the majority of the scientific establishment but not really "an aether" - a Higgs field filled with Higgs bosons, which is itself embedded in a matrix of dark matter [23% of the universe they say], or vice versa, which is then embedded in an even more dominant matrix of dark energy [73% of the universe], and all this is part of the space-time continuum! All three of these have been invented to explain the phenomenon for which there was no explanation based on our current most accepted models. These entities have replaced the aether with a level of complexity that seems absurd, and yet they still cannot explain many aspects of physics by invoking these replacements for an aether. The Higgs mechanism arising from the inability to explain mass, dark matter from the lack of understanding the flat velocity profile of the motion of stars of the galaxies, unusually large groups of galaxies, and dark energy to explain the inflation of the universe which is entirely based on the redshift of light from ever more distant galaxies.

The new model would have to be able to literally explain everything that anybody could think of. Which I realized was more than reasonable if the theory was so good then it should be able to explain the phenomenon that other theories could not account for, and not simply explain away something by saying some magical particle generates that force or effect. It needed to explain everything without such an excuse of simply pushing off the explanation to some other particle or field. And without reverting to a mathematical explanation that few could understand. Which was no problem at all for me since my own algebraic, and calculus, skills were limited to only a few variables. I'd have no choice but to explore the physics of physical processes based on known experimental data, and phenomenon.

If Sir Isaac Newton [1642-1727] had known what we now know about some of the empirical data around nuclear chemistry and radioactivity, we would likely not be having this discussion as he can be credited with what appears to be the true nature of what gravity can do, and with such knowledge, he, or someone else, in time would have come up with the same model that is presented to you in these pages.

Where does this revised model of an aether point us towards?

It points us towards a model of the universe that has - No time travel, No inter-dimensional travel, No parallel worlds, No inter- dimensional beings visiting us from hidden dimensions, No faster-than-light drives, No tachyon faster-than-light communications, No multiverse, No wormholes, Not even black holes — see homopolar galactic generator model, and yes galaxy scale electric currents and magnetic fields even take into account the Sagittarius A* stars at the core of our galaxy and their motion about its center. Electrical and plasma engineers predicted double active-cores at the centers of galaxies before astronomers observed pairs of something happening within them. Charged bodies [ionized gas giants called stars] are electrically propelled about the center of all spiral galaxies. If there was a black hole at the center of our galaxy not only would the Sagittarius A* stars move about the core of our galaxy rapidly but there should also be some optical distortion as they pass between the black hole and us. The homopolar galactic generator model not only accounts for this better, but it also long ago accounted for the detectable gaseous magnetic tendrils observed at the core of our galaxy, and the flat velocity profile of stars around galaxies — that dark matter was conjured up to explain. Black holes that finally get so big that they finally emit astronomically massive quantities of energy, due to the compression of matter just before they pass the event horizon, are a contradiction of the very definition of black holes, and the electric model is so much simpler. *Now they are claiming to see a greater frequency of pairs of black holes in several galaxies, again the electric model predicted pairs of electrical "cores" like this during the merging of galactic nuclei.

Again, what it all seems to imply - No extra dimensions, No time travel, No black holes, No wormholes, No other universes inside black holes, No singularities, No dark matter, No dark energy, No anti-matter anomaly, No inter-dimensional beings, No multiverse, No quarks, No gluons, No gravitons, No monopoles*, No super-strings, No supersymmetry, No sparticles, No branes, ...

A reasonable explanation for gravity, quantum mechanics, electron orbital positions, photoelectric effect, Observer effect, inertia/mass, two slit experiments, Zeeman effect, Stark effect, radioactivity, Electron Capture, wave versus particle duality, color/strong force, electro-weak force, spectral lines, what makes some atoms a gas, the Bell experiments, "action at a distance", why is mass restored after fission,...

What did Einstein think...
"You imagine that I look back on my life's work with calm satisfaction. But from nearby it looks quite different. There is not a single concept of which I am convinced that it will stand firm, and I feel uncertain whether I am in general on the right track."
-Albert Einstein

Gedankenexperiment

The effect that black holes are supposed to have upon the paths of photons, who venture too near to them [and the fact that hydrogen atoms are not chemically neutral, and why neutrons were not stable outside the nucleus] is what triggered my inescapable quest for a mechanical theory of the nature and structure of the universe. My eureka moment was when I was doing a thought experiment, gedankenexperiment, of following the paths of photons as they passed by a black hole. If a photon could be drawn into a black hole by gravity, away from the path it would of otherwise have taken, then the simplest physical explanation could only be that such waves seemed to be drawn to the space around a black hole as if it were becoming an increasingly more effective conductor to the photons. [Like an electron's attraction to a positively charged ion or atom.] The only physical means of conduction that I could think of was if a black hole was enveloped by a fabric of space (FOS) of an ever increasingly greater density[9] nature. The gravitational field has been portrayed as a distortion in space-time itself, but equivalently Einstein's equations can also be portrayed simply as a density gradient, in/[of] the fabric of space. [The density gradient is the gravitational inducing field. The same gradient that Newton was pondering and that gives the appearance of spontaneous action at a distance. And the same gradient developed for and implied by Einstein's curvature of space equations. Remember they are still based upon the gravitational constant, which is the factor that is based upon the gravitational attraction of any number of bodies that are charge-neutral masses.] According to Einstein, as he is often cited, gravity isn't a force acting between masses but due to the curving of space-time triggered by their presence. So, as the Moon travels around the Earth, it is in fact following the contours of the space-time distortion in space. Often portrayed by a marble rolling around the distortion created by the presence of a larger mass on a stretched sheet of rubber.

---

[9]the ratio of a mediums rigidity vs. its density controls a waves velocity of conduction.

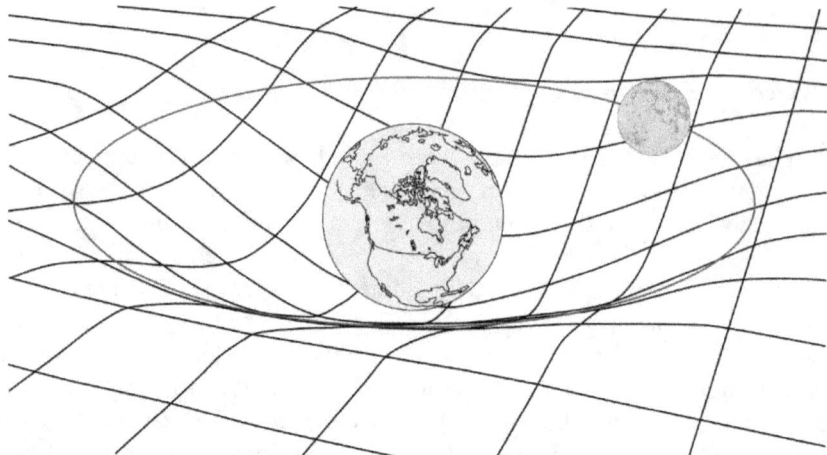

Figure 23- Distortion in space-time depicted by balls on a rubber sheet

Regarding this envelope of material around any mass made me think that the waveform that most easily fit the necessary requirements for conduction along this domain of reasoning was that of a longitudinal wave. But of course, photons are supposed to consist of transverse waves. Why? Because only transverse waves are supposed to be polarizable, and work their way around hard corners like water waves versus a spray of particles. [Which is what we just covered in part one of chapter one.] And yet transverse waves are shown to be so much slower than longitudinal waves when they're within the same medium. Think of the waves on the surface of water versus the sound waves passing through the water. And then there's the Compton Effect of x-rays interacting with electrons*.

The problem is that most people associate longitudinal waves only with sound waves. Consider how fast sound waves seem to decay or dissipate. I am not suggesting that we consider a sound wave as an exact model for a photon, but they do share some key characteristics. The phonon and a photon would seem to have a similar principle of transmission, but a different overall reaction sequence of dispersal versus adherence over time. A phonon, whose energy is carried by a high-density accumulation of energetic atoms, very quickly spreads out and disperses the energy it contains partly due to the spaces between the atoms in a gas. A lot like billiard balls that collide with each other and that can pass by one another. Consider the size of atoms and the average space between them when they are in the form of a gas. Now consider the opposite, if an aether exists its scale of granularity must be far greater in magnitude. If you compare the diameter of the hydrogen atom to the diameter of its nucleus this change in magnitude is on the order of something like 145,000 times smaller. We have to consider what some think Planck's constant implies. Some think it implies the size difference to be of a much greater magnitude of finer material. Some say it implies a 10^{-33} level of granularity. [Versus the size of the hydrogen atom or the unit of length the metre?] For the photon consider basically a focused pulse generating a wave-front of this aether undergoing compression and rarefaction, but much more like a semi-solid in terms of density rather than gas which has vast distances between atoms. With the wave front being roughly circular in nature, but not forming a perfect circle. And this is key. Perfect geometric forms are more of an ideal rather than a common occurrence in nature. If a longitudinal wave has an elliptical profile, then it too can be polarized. The more elliptical vs circular, the more readily they can be polarized. Add to this the patterns generated by the two-slit gap diffraction experiments, along with the observer-interference effect, and it is easy to see that de Broglie [&Deutsch] was correct in his hypothesis of the existence of pilot waves. Pre-compression waves forming ahead of the main body of a photon. An experiment might show differences in the polarizability between different frequencies of waves. Red light versus x-rays. Which is true if you look at how difficult it is to polarize x-rays.

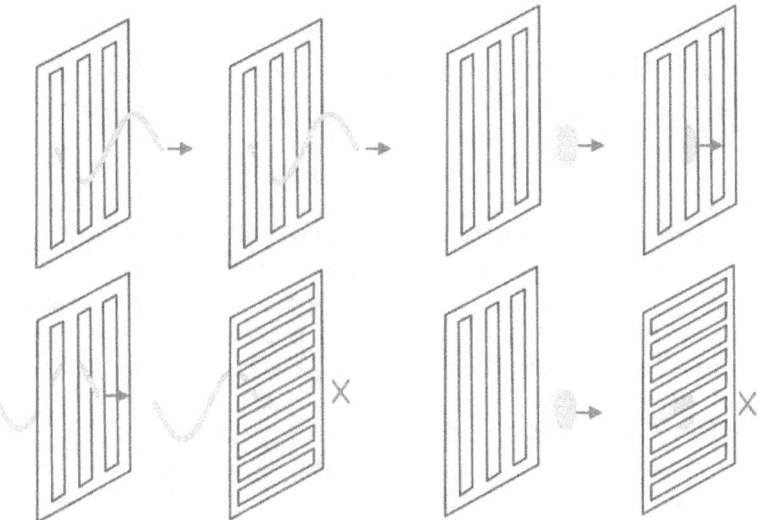

Figure 24- Both transverse waves and longitudinal waves can be polarizable

Pilot waves are missing from this image. They help photons to twist through a middle filter at 45 degrees to the others, as this supporting higher density material bends around the polarizing filter's structure.

Transverse waves are often portrayed as two-dimensional sinusoidal waves and yet photons behave in experiments as if they are in fact three-dimensional entities. If you look at the Compton Effect where electrons are interacting with x-rays, the x-rays appear to behave more like semi-solid masses and can lose [penetrating] energy while the electron gains kinetic energy or vice versa.

$$E_k = E(x\text{-ray}) \, , \, QV = hf_{max}, \, E(x\text{-ray}) = hf_{max} = h(c/\lambda)$$

$\lambda_{min} = hc/(QV) = [(6.63 \times 10^{-34} J.s)(3.00 \times 10^8 m/s)]/$
$[(1.60 \times 10^{-19} C)(2.3 \times 10^4 V)]$
$\lambda_{min} = 5.4 \times 10^{-11}$ metres for x-rays whose E = 23,000 eV
5.4×10^{-7} metres for green light whose E = 2.3 eV

Most x-rays generated by laboratory equipment would be less than the maximum value. Typical voltages used to generate x-rays vary from 23,000 volts to 27,000 volts.

X-ray machines for deeper imaging using 30,000 to 100,000 volts.

Linear accelerators can be used for x-ray therapy with voltages of around 6,000,000 volts. Arthur Compton looked at x-rays entering matter which released weaker x-rays and from studying them discovered that the total momentum, as well as energy, is conserved. Which proved that they can act as both waves and particles.

$P = mv$, $E = mc^2$, $p = (\frac{E}{c^2})V = (\frac{E}{c^2})c = (\frac{hf}{c}) = (\frac{h}{\lambda})$

De Broglie carried this further to show that electrons can have a wave length.

For the Compton Effect it makes more sense to view them as a compressional wave such that the high-density body is transferring energy to the electrons and having its density reduced in the process. An explanation of the Compton Effect involving transverse waves is not at all obvious using a mechanical explanation. Remember that the UV catastrophe and the initial expectations of a "slowly increasing" photoelectric effect based on the transverse wave model completely failed in its predictions.

Another interesting photonic effect is related to the polarizing of photons and how they can be forced to change their polar alignment, by passing them through a second linear polarizing filter, at roughly 45 degrees in orientation to the first one, on their way to a third filter which then allows some of the photons to pass through the third filter which was at 90 degrees to the first. Without this middle filter, no light passes through the third one, and by adding this middle one now, some light pass through the third.

The Death of the Dark Energy Idea

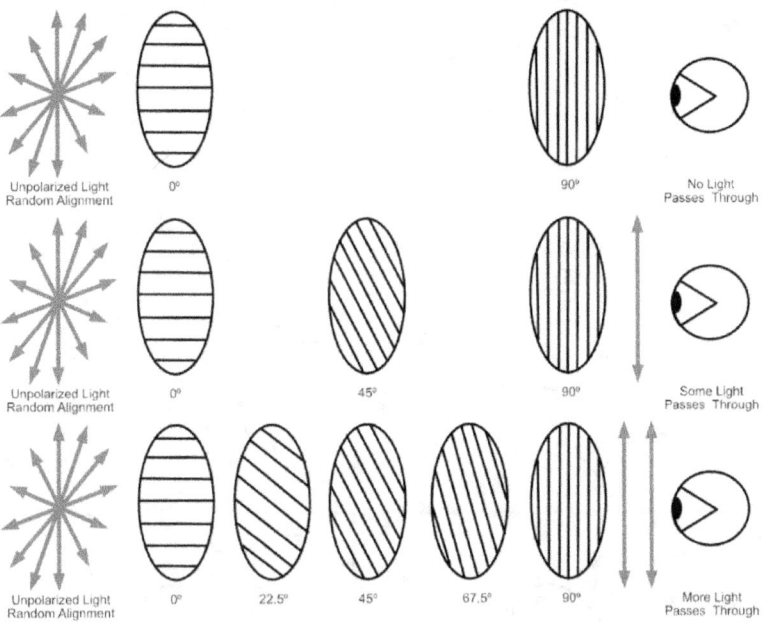

Figure 25 The more polarizing filters the more light passes through.

Sinusoidal 2D transverse waves do not suggest any mechanism in which they can be forced to rotate to pass through a polarizing filter that is not aligned with the first one it has passed through. Not even in the case of a quarter-wave retarder systems where photons are not just passing through linear slits, but instead pass through a material which triggers a planar orientation change. This behavior implies something else as well and would seem to indicate a lower density form being generated ahead of the densest part of photons. But which also has the ability to guide the photon if this material/region is affected enough by something ahead of the photon. The implication of this has a bearing on the effect related to photon two-slit experiments as well as for electrons.

In the next two figures we can see that when a photon, preceded by its pilot wave, approaches a regular barrier the bounce back is of a uniform density gradient nature. But when it is near either a gap in a barrier, or the edge of some obstacle, then we end up with at least one plane of material that is of a higher density. With such a planar region ahead of it then our conducting photons' mass aligns itself as best it can over the given distance ahead of it. Literally twisting during this change as it adapts / conducts along the better gradient region. This means higher rates of polarization and accounts for why more photons can pass through a series of polarizing filters. No mysterious quantum mechanics involved.

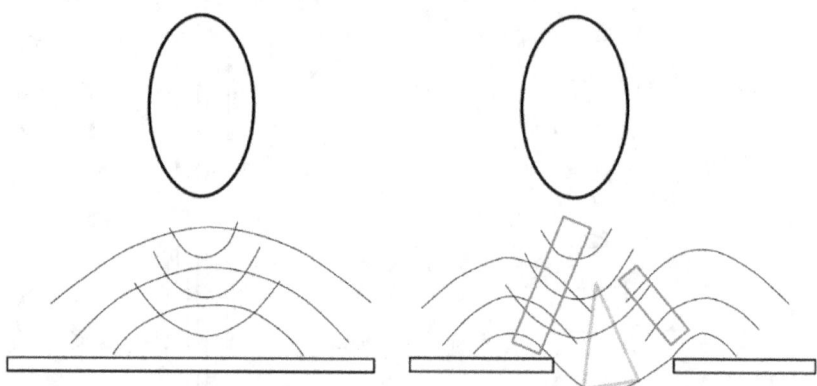

Figure 26 Pilot wave reflection near an edge leads to a higher density plane.

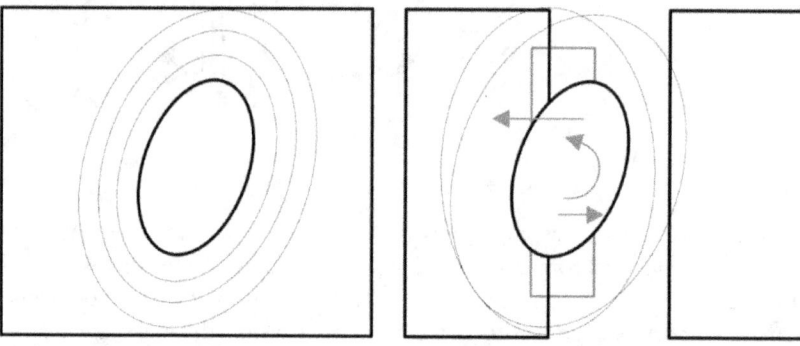

Figure 27 Pilot waves induce both photons and electrons to align with the more conductive path.

Figures 26 and 27 show how pilot waves are affected by objects in their way, and how their conductive bodies affect photons into aligning as best they can to these more conductive paths. Triggering their rotation, to some extent, and then being able to increase their odds of passing through polarizing filters.

This behavior triggered by the formation of what are called pilot waves is stirring up some excitement on the internet and in science shows. It was Louis de Broglie who first proposed the idea of pilot waves travelling along with electrons to give them their wave-particle duality, and later championed by Sid Deutsch. In [2010*] Yves Couder and Emmanuel Fort from the University of Paris performed some astounding experiments that hint at what might be behind some aspects of quantum mechanics. Louis de Broglie [& Sid Deutsch] brought up the idea of pilot waves as a mechanism to account for some of the unusual effects seen in quantum mechanics. Couder and Fort have shown the power of pilot waves in experiments with small droplets of silicone suspended over a vibrating fluid mimicking quantum effects like that seen in two-slit experiments. Videos of their work have been posted on the internet for a few years now.

Photonic shape, at least around the visual part of the spectrum, can be imagined as half of an ellipsoid. A rather simplistic idea of something more complex. But with an elliptical shape for their frontal profile, they then can have literally both height and width. Thus they can be polarized, and like all waves still possess all the other characteristics (reflection, refraction & interference patterns) that experiments have shown them to have. One must remember here that wavelength would be more accurately described as wave spacing or literally length, and not a physical characteristic denoting frontal width or height. What about circular polarization and photon angular momentum? What else could this imply? Also, what about the movement of the pilot wave material as it moves out of the way of the passing photon? Does this motion contribute to the magnetic field being generated, or is it strictly angular momentum?

Polarizable longitudinal waves? There should be some kind of a link between the equation for longitudinal waves and photons. And there is with the FOS aether model. Over the years that this model was developed, I had not realized that my approach to how electrons and magnetism hinted at working would lead back onto itself and provide evidence for itself through the model and the constants for electromagnetism. Which in turn provided additional proof of the correctness of the FOS model for photons. The nature of electrons' effect on space was being teased out of the data.

Given that the decay products of the matter/antimatter annihilation event of an electron-positron pair produces two gamma-rays, sometimes three, consider then that the origin of an electron is a gamma-ray, and as such its behavior should be related to the same components that affect the conduction of a photon. If a photon is a longitudinal wave, then one should be able to link the equation for longitudinal waves in matter to gamma-rays or at least to one of their alter egos. Like an electron.

$$\frac{\textit{The Bulk Modulus of Elasticity (Of some medium.)}}{\textit{The Density (Of some medium)}} = \textit{The square of the velocity of a wave.}$$

Or $\frac{B}{p} = v^2$

From Wikipedia for Bulk Modulus: "The bulk modulus (K or B) of a substance is measure of how compressible that substance is. It is defined as the ratio of the infinitesimal pressure increase to the resulting relative decrease of the volume."

The electrical force constant [from Wikipedia]
$K_e = 8.9875 \times 10^9$ N·m^2 / C^2 or kg·m^3/s^2/C^2
$$K_e = \frac{1}{4\pi\epsilon}$$

The magnetic force constant [from Wikipedia] [error in characters]
$K_m = 4\pi \times 10^{-7}$ N / A^2 or N·s^2 / C^2 or (kg·m/s^2)·s^2 / C^2 or kg·m/C^2 ≈ 1.2566370614...×10^{-6} H / m or T·m / A or Wb / (A·m) or V·s / (A·m)

If one takes the Electrical Force Constant and divides it by the Magnetic Force Constant, we end up with the velocity of light squared. In other words, if one takes the factor of the FOS that determines the strength-rate by which electrically charged bodies can either come together, or move away from one another, and divide this by the factor of the FOS structure that determines the strength - rate by which two magnets [or two magnetic bodies] can come together, or move away from one another, - we then end up with the velocity of light squared.

$$\frac{Ke}{Km} = \frac{kg \cdot m^3/s^2/C^2}{kg \cdot m/C^2} = m^2/s^2 = c^2$$
Where C is Coulombs, and c is the speed of light.
gravitational constant = Kg = 6.67408 × 10^{-11} m^3 kg^{-1} s^{-2}

It was the proof that linked electricity, magnetism and photons together. It is an already well-accepted fact that an electron–positron pair can be formed from the interaction of two gamma rays. An odd thought that we can go from what gamma-ray waves to matter. How? Part of the answer is in how we define what matter is.

By comparing the factors involved in longitudinal wave propagation with the characteristics of the FOS that directly controls, and affects electricity & magnetism, it can be seen that they appear to be related. The bulk modulus of elasticity is defined as being the "compressive (or tensile) force per unit area divided by the change in volume per unit volume." And normally of course, density is defined as the mass per unit volume or as the inertial property of the medium. In the model, we are about to present to the reader the above equation is far more relevant and surprising in its principle* of equivalence. In this new FOS model, the key factor involved in electrical attraction is based upon how the passage of an electron through a region of the fabric of space, and how it disrupts and decreases the normal density of the fabric (change in volume per unit volume), which then sets up a potential difference in the pressure of the surrounding fabric. Which permits the fabric outside this immediate region to force the fabric, by pressure, back into this space to normalize the density gradient & pressure. This mechanism permits electrons to force protons to remain in proximity with one another and help form the nucleus of an atom.

In magnetic attraction, one of the key factors in the formation & existence of a magnetic field is how well the parent electrons can keep the fabric in motion, or at least some inertial component appears to be involved. This is in relation to the resistance of the fabric to being set into motion (an inertial property and one related to mass) and how easily it can be maintained against the local fabric from absorbing this energy. (What role does the momentum of the aether play?)
So, in the FOS theory series, there is a correlation between the bulk modulus of elasticity for longitudinal wave propagation via how the fabric pressure per unit area affects the negative change in volume per unit volume. While correspondingly, density in the equation for a longitudinal wave is paralleled by the momentum or inertial property of the fabric that controls magnetic fields within the fabric of space. Thus, we have an equivalent relationship between photons and phonons with respect to the square of their potential velocity.
Michelson & Morley proved that the Earth was not traveling through an aether with any measurable drag factor, which flowed through or around the surface of the Earth. They were trapped, and took almost everyone else with them, into one mindset of how they thought the universe should work and how the Earth should move through the aether. And their inherited concept of the Earth composed of solid atoms is one of the key ideas about why they were wrong. They first performed their experiments in 1887 when our idea of atoms was that they were solid objects. An idea promoted by John Dalton, who had decades earlier [1803] proposed his Atomic Theory of Matter where atoms cannot be subdivided, created or destroyed. They were locked into the concept of the Earth as a solid body that was composed of solid atoms. It was not until 1911 when Ernest Rutherford [et al.] showed the world that atoms were mostly empty space. Regardless of the practical radius you choose for the hydrogen atom, empirical versus Bohr, versus the size of a single proton, the average hydrogen atom is about 99.9999999999996% empty space! Basically nothing! If I told you I had removed 99.99 percent of something you would think I had nearly eliminated something. If I told you I had removed 99.9999999999996% of something – could there possibly be anything left?! This shows us how small the nucleus is compared to the size of the hydrogen atom. Today a number of accepted experiments do show evidence of some degree of the twisting of space around the Earth [aka frame dragging]. This is likely the shearing/deformation of the FOS gradient formed around the Earth, but might also be an effect related to the Earth's pilot wave. The FOS model proposes an aether, unlike the modern hybrid mixture of a Higgs field mixed with dark matter and dark energy. With the FOS at its lowest density between bodies of matter, and at its

highest density around bodies of matter. Reaching its peak of greatest density at approximately near the surface of nuclei/protons.

What appears to have happened is that M & M were searching for a flowing aether. In their day it was believed that matter was made up of solid material that then obviously the aether would be like a wind flowing over its surface. As we argued in part 1 of chapter 1 is that their work is a moot point as we already pointed out that a moving train does not change the speed of sound, and can only change the frequency of sound. Their basic premise of c+v or c-v is not valid for a longitudinal wave within a conductive medium. In a universe that is really composed mostly of a truly empty space, then photons would not be limited to a mediums ability to conduct them. Even transverse waves are still boundary wave distortions of something, and as I previously stated, do not apply to three-dimensional space. While with longitudinal sound waves a moving train can only result in changing the density of the waves during formation and thereby either add to their frequency or decrease their frequency. Meanwhile, the fabric of space that surrounded Michelson & Morley was of the same density regardless in which direction they faced. The FOS surrounding them was in ever-decreasing layers, moving away from the earth, of density - spheres around spheres until they blend in with Sun's gradient than the rest of the universe. The analogy that comes to my mind is that M & M's grand experiment was similar in nature to two divers trying to detect a variation in the pressure of water at one depth. The critical idea here is the density. It is what causes a photon that passes too close to a star to bend its path noticeably. Einstein's gravitational field description is no more than that. A description of how matter and energy respond to large concentrations of matter. It is simultaneously a description of a gradient and is so because it is based upon the gravitational constant of the attraction between charge-neutral masses.

These layers are of a decreasing density nature, which helps to explain the variable gravitational acceleration rate that is inversely proportional to the square of the distance from the Earth – $1/R^2$. The acceleration rate is a product of the FOS density gradient and electrically neutral matters' interaction with it. In other words, the rate of change in the density of the surrounding fabric is the cause of the rate of change in the acceleration of an electrically neutral body to the center of mass.

The point of the FOS model is to describe mechanically the how's and why's of atomic physics. And by making the minimum number of physically unexplainable suppositions and minimize the number of fundamental forces (gravity, electroweak & the strong nuclear forces are just derivatives of the electromagnetic force), and avoid implying effects for which little if any physical/mechanical explanation can be given.

A photon, in this model, is composed of a longitudinal density wave of FOS. But a critical density exists at which, if achieved, through a directionally focused pulse, the body can exist practically indefinitely. But not forever, as the extreme redshift of the most distant galaxies shows us. The greater the initial density of a photon, the longer it will take to decay. Thus, gamma rays last the longest and radio waves decay the fastest. Not that their decay rates are different it's just that radio waves are already weaker and the various photons often interact in different ways with material and electromagnetic fields that exist between the stars. [Not sure if I am making my intended point.]

When a photon travels across a density gradient, it sees the FOS closer to the mass, producing the effect known as gravity, as a better conductor. And as a result, is drawn towards it because the gradient induces into the conduction of the photon a lateral drift component. Resulting in the motion of the photon shifting, to a small degree, towards the mass. The path being altered to the degree that is proportional to the intensity and size of the FOS gradient of the gravitational mass. In other words, a near-uniform gradient produces a negligible change in the path of a photon. Whereas if there is a greater change in the FOS density gradient, then a more noticeable change in the path of the photon may be detected.

Basically, gravity induces something akin to "refraction" in the photons that pass through such a field like that of our theoretical black hole. Both Newton's and Einstein's work predicted this but Einstein's work was more accurate. If you're up on your physics you might try to point out that according to Snell's law of refraction that the angle at which a beam of light bends into a transparent body is believed to be based upon how much the material of the body slows down the beam of light. [Interesting how this theory allows for the violation of the conservation of the speed of light, and Einstein's statement that it is constant.] The less the photons are slowed down, the smaller their angle of refraction;-the less the beam, that they make up, bends. The reason I brought this up is that if you knew Snell's law then you would probably have disagreed with my next point. That is if you thought that I was not aware of it. The passage of light through a transparent body is not the same as the passage of it through the vacuum of space. One only has to look at the passage of photons through a prism to realize that there is more to the conductance of photons through bodies of matter. I will argue that they are not based on a similar principle on the basis that, if the photons were slowing down when they encountered and entered into a gravitational field then they would become red-shifted and not blue-shifted as experiments (see the measurement of the gravitational redshift of photons using the Mössbauer effect) has shown them to become (i.e. more energetic). And at the same time for the sake of evidence, it is well known that light becomes noticeably red-shifted [increases in wavelength] when it is emitted from the surface of a very massive star. Photons see the gravitational field [density gradient] as a better conductor, and thus their bodies are conducted deeper into these regions. This idea is key to how electrons behave as well.

Gravity as the outcome of an aether density gradient? Sir Isaac Newton was the first to consider this as a result of his mathematical exploration of gravity. Experiments are detecting something around the Earth. Even if we can only state that they are detecting the twisting of space as the Earth rotates on its axis. This "something" is being detected by measurement of the "twisting of space" around the Earth by Ignazio Ciufolini [et al.][1984], Gravity Probe B[2004?] and Roland De Witte[1991] [aka frame dragging]. I have pondered that this could be due to the Earth's own pilot wave.

Sid Deutsch, Louis de Broglie and the Aether

It was Sid Deutsch (Deutsch 1999) who answered a couple of questions of mine that had me perplexed. I was asking myself what is the implication of the FOS/aether build up in front of an electron after its formation from a two-photon physics event. It was he who introduced me to the concept of pilot waves, or what Erwin Schrödinger mockingly called ghost waves(Gespensterfelder) as Alberta Einstein toyed with the idea of them. Sid Deutsch did not mention or seem to be aware that Louis de Broglie appears to have been the first to set the ground work for them for photons and matter. While it was years earlier that Albert Einstein considered ghost waves for guiding the photons. As opposed to simply preceding the main mass of a photon as it enters a region of space, but still affecting the behavior of photons under the right set of conditions.

Einstein's view of the aether can be summed up as simply as – "That is that you could simply ignore the question of the existence of a luminiferous aether. Forget trying to explain it, and just accept the fact that the speed of light is a constant." A constant velocity of light is in fact, evidence itself of an aether. Since it implies an aether restricting and setting one and only one velocity for waves. [M&M proved no entrained aether for a solid atom Earth.] The sound-based Doppler experiments for some vehicle [usually a train, police car or ambulance] moving through the air near you tell us what we should expect to see. The velocity of the sound wave does not change; instead, its pitch changes at first to a higher compression as it approaches you then to a lower compression as it moves away from you. The speed of sound does not change. This effect is observed in astronomy as well and tells us whether, relatively speaking, if we are moving closer to another star or away from it. This is seen as the famous motion-induced red or blue shift of the light. It is the apparent redshift of light that is believed to indicate the expansion of the universe and evidence for the big bang theory of the known universe's start. Which you now know is simply an erroneous interpretation of relying on a transverse wave model for photons versus the simply mechanical expansion of photons based on a longitudinal wave photon model.

As a reminder, the Special Theory of Relativity, Einstein's extension of the work of Kepler, Galileo and Newton, can be summed up as that the laws of physics are the same for all observers in uniform motion, and the speed of light is a constant. And this has nothing to do with $E = \text{mc}^2$ or his work on the Photo-electric Effect or Brownian motion. While in General Relativity, the laws of physics are the same for all observers regardless of their motion. And that gravitational mass and inertial mass are equivalent. Depending on which of his quotes you use or reference you can have Einstein say pretty much anything you want about the aether. But ultimately, he simply ignored it and wanted the topic to go away. In his time Dayton Miller's work was an annoyance to him, and he was quoted as saying: "My opinion about [Dayton] Miller's experiments [once thought to prove an entrained aether [not supported by the FOS model]] is the following. ... Should the positive result be confirmed, then the special theory of relativity and with it the general theory of relativity, in its current form, would be invalid. Experimentum summus judex. Only the equivalence of inertia and gravitation would remain, however, they would have to lead to a significantly different theory."

— Albert Einstein, in a letter to Edwin E. Slosson, 8 July 1925

Invalid in its current form? The basic principles of relativity, both special and general, would not be wrong with an aether. The laws of physics would still be the same regardless of the motion below the speed of light, and an aether would enforce the constancy of the speed of light. Not give it latitude to vary significantly. However, the time issue arises, and that is different and in opposition to the FOS model. But as we will see the density and density changes can account for the time dilation effects we observe without invoking magic; this mass increase simply effects the resonant frequency of atomic clocks. Along with muons who are raining down on the Earth having their decay slowed possibly because they feel the back pressure on their form as they enter the Earth's density gradient. Or could it instead be that they see the Earth's gradient as more conductive and thus are drawn in by conduction and thus their velocity increase? A measure of their velocity might clarify this but could their increased momentum confuse the two measurements?

Regardless of the experiments, we can see that something is seen as twisting as the Earth rotates about its axis. [aka frame dragging] Or the Earth's pilot wave is triggering the measurable effects.

[Need to cover more of Deutsch's work but also the twisting of space.] [aka frame dragging.]

All aether proponents point out that space has measurable properties, not something you expect if there is "nothing" to have properties.

[More to come.]

The Fabric of Space versus a Higgs field + Dark Energy

Newton's ideas about space, time and motion seemed to imply that the speed of light would have different values for objects and experiments that compared themselves to the perceived speed of light for non-moving objects and experiments. No experiments supported his idea of the relative speed of light. He was one of the first to support Descartes theory of a corpuscular theory of light. Michelson and Morley could not detect any difference with their experiments, while Dayton Miller's ether-drift experiments did seem to indicate a difference. Why? With Miller's experiments being at different altitudes and in an open environment as opposed to M&M protected basement experiments. Could he have in fact detected frame dragging or as I have implied density gradient sheering or the Earth's pilot wave triggering what we think of as frame dragging? [Need to investigate this further.] In the end, the matter of the problem was silenced when Einstein proposed that everyone needs just to accept the fact that the velocity of the speed of light is a constant. End of story. No. Ironically the very notion that the speed of light being a constant is an indicator of a luminiferous aether. And we can detect relative variations between bodies and it is spoken about all the time but without linking the two together. It is called the redshift of light. When the speed of light is compared between two objects moving at significant speeds relative to one another, the presence or absence of the redshift or blue shift of their light sources is well known and understood. Today we even take this notion of an absolute frame by comparing our motion to the cosmic microwave background radiation.

One of the problems that Einstein raised was that Newton's law of gravity seemed to infer a much a faster consequence of the influence of gravity than could be imposed by a limit on the speed of light. This anomaly of course actually being evidence and support for a density gradient that exists at all times, in effect travelling as the Sun moves, and thus the motion of the planets is not limited by the speed of light upon some form of gravitational waves acting between objects. Einstein resolved this with his warping of space and time. The FOS model within this book will show that Einstein's equations and ideas can be simplified to a simple density gradient of an aether, and waves within it cover everything else as well. Versus a Higgs field, dark matter and dark energy matrix that is in the end just a much more complex aether and still requires a whole lot of exotic particles and the distinct four [three] forces to cover it all. Which does Occam's razor lean toward?

Tired light [TL] theories are stuck on the concept of transverse waves, for which no one ever explains how a restoring force would work to keep the waves in motion, and often talk about the loss of such of energy by such photons showing up as the redshift we observe. This redshift, in tired light theories, is supposed to occur due to the loss of energy by the photons because the photons have lost small amounts of energy as they pass close to, or through, molecules and particles in space. Halton Arp pointed out that no additional redshift is seen when we look through our galaxy in the direction of relatively high-density clouds of gas and dust as compared to regions that look about the same distance that do not have such clouds. Astronomers do see clouds of hydrogen gas absorbing light and redirecting a number of the photons passing through them and showing up in the form of an absorption spectrum. Still, they do not observe any additional redshift as a result of their passage through the gas clouds. Thus, the most popular tired light theories are not supported. In the FOS model, we are not looking at or even considering magically propagating transverse waves, instead, we are looking simply at compression [longitudinal] waves of the fabric of space itself. The redshift is occurring due to the simplest reason – they are simply spreading out. Crudely speaking like how the waves on a pond spread out. [Is there any evidence of phonons, sound waves, reducing in frequency over long distances? Not related to friction.]

The amplitude changes of transverse waves

Transverse waves, which are boundary waves, change over time. And one reason is dampening, and friction is one of the components of this decay. But more important is the transmission dynamics that is the main cause that reduces the wave's potential energy over time.

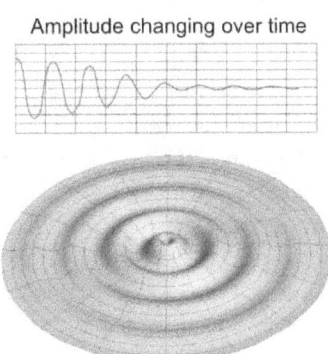

The reason is that as the waves propagate outward, their energy is spread out over a larger area/volume. Similarly, sound waves spread out as they propagate. But are often more focused during their formation and thus travel further before decaying in part because of this focusing.

Figure 28- Transverse wave amplitude reduction

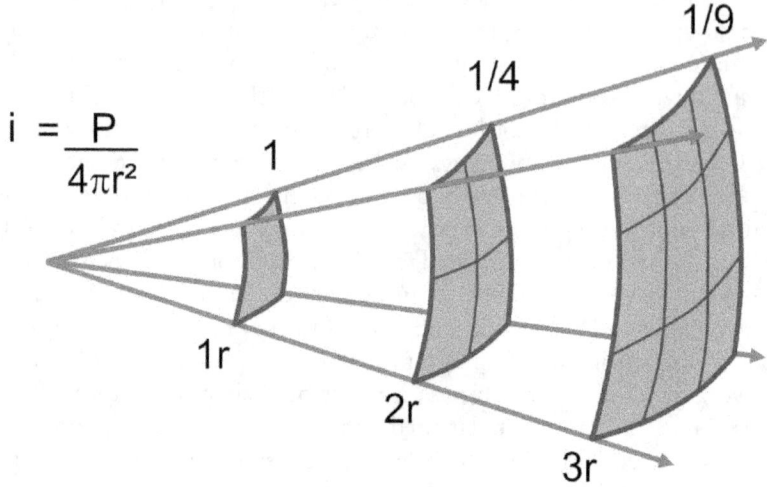

Figure 29- Inverse Square law for sound wave intensity reduction

Other Considerations

They still wish to treat the universe as something without a conductor that even in the absence of any large masses still controls the speed of light and the motion of atoms, ions and electrons in the absence of a gravitational field generated by some significant-enough of a mass. Many claim the universe has no center and yet next say that it is expanding outward. They are inevitably assigning it a center and an age based on that boundary. Thus, they create in their minds the need to apply a center of mass to the universe, and neglect that there can be motion without that center. It does not make sense to talk about a perfect theory that relies heavily on the gravitational constant, and with it only considers neutrally charged matter, and masses, while ignoring the electrical and magnetic fields that permeate the universe. In a universe filled with electrically active stars and galactic nuclei. [More to come.]

This so-called perfect theory relies on the electrically-neutral gravitational constant as its foundation, and because of this fails to predict the flat rotation curves of the motions of the stars within galaxies. This failure eventually giving rise to the theory of dark matter to account for this problem. [Eventually it was used to justify the perceived expansion of the universe by reusing the universal constant.] A universe originating from a single point that comes about because of the big bang, whose existence is entirely based on the redshift of most distant stellar objects that we can detect. Part of the problem is that gravitational measurements are based on a center of mass, even when there is no center.

The posited aether is modeled substantially on the form of aether described by James Clerk Maxwell in his treatise, *The Dynamical Theory of the Electromagnetic Field*, published in 1865. It is in this treatise that Maxwell set out his theory of electromagnetic fields and his equations that are still in use today, over a century-and-a half later. His theory and equations were based upon his posited form of aether[10]. In the words of Maxwell:

- "We have therefore some reason to believe, from the phenomena of light and heat, that there is an ethereal medium filling space and permeating bodies, capable of being set in motion and of transmitting that motion from one part to another, and of communicating that motion to gross matter so as to heat it and affect it in various ways."
- "We may therefore receive, as a datum derived from a branch of science independent of that with which we have to deal, the existence of a pervading medium, of small but real density, capable of being set in motion, and of transmitting motion from one part to another with great, but not infinite, velocity."
- "Here, then, we perceive another effect of electromotive force, namely, electric displacement, which according to our theory is a kind of elastic yielding to the action of the force, similar to that which takes place in structures and machines owing to the want of perfect rigidity of the connexions."
- "It appears therefore that certain phenomena in electricity and magnetism lead to the same conclusion as those of optics, namely, that there is an ethereal medium pervading all bodies, and modified only in degree by their presence; that the parts of this medium are capable of being set in motion by electric currents and magnets; that this motion is communicated from one part of the medium to another by forces arising from the connections of those parts; that under the action of these forces there is a certain yielding depending on the elasticity of these connections; and that therefore energy in two different forms may exist in the medium, the one form being the actual energy of motion of its parts, and the other being the potential energy stored up in the connections, in virtue of their elasticity."
- "Thus, then, we are led to the conception of a complicated mechanism capable of a vast variety of motion, but at the same time so connected that the motion of one part depends, according to definite relations, on the motion of other parts, these motions being communicated by forces arising from the relative displacement of the connected parts, in virtue of their elasticity."

---

[10] From Duncan Shaw on Maxwell's work

The Death of the Dark Energy Idea

Chapter Three

THE ELECTRON

The universe is comprehensible because it is fundamentally simple in nature. This is why the mathematical relationships we've teased out of the data consistently works in the first place as a reflection of this simplicity.
— Terrance J Fidler

A theoretical construction is unlikely to be true, unless it is logically very simple.
— Albert Einstein (1879-1955)

The Electron

Although we have introduced the idea that a FOS density gradient is in effect the gravitational field, no explanation for the affect it has on matter has yet been given. I had suspected that the origin of the force must lay within the structure of the electron. In search of a clue, I turned to the relationship between electrons, positrons and gamma-rays, and on the fact that they can be derived from one another. This is found in the annihilation reaction between the electron, and its' anti-particle, the positron, and results in two (sometimes three) gamma-rays (and sometimes a neutrino is detected). The reverse reaction called two-photon physics is where the interaction of two gamma-rays results in the production of an electron and a positron. Another idea that came to my mind to support, and consider, in a proposal for a new model for the electron was the particle-like behavior in collisions with x-rays in what is called the Compton Effect. The Compton Effect is the transfer of energy from x-rays to electrons. The electrons gain velocity and the x-rays increase in wavelength. Although when I came up with a possible structure for an electron, I was unaware of a lot of known physical phenomenon and observations, but I went ahead and based my ideas on the concept that bodies within the fabric of space were based upon pockets of the fabric varying in density, and with all of space possessing a plastic-like property granted to it by the material that it is composed of. [The Compton Effect clearly indicates particle-like properties of not just electrons but also of high-energy photons like x-rays and gamma rays. We will look at the simple nature of wave-particle duality soon.] What this effect consists of is the observation that in the collision between an electron and an x-ray that the electron gains kinetic energy from the collision, as if two billiard balls had struck one another, with the x-ray increasing in wavelength (becomes less dense). Clearly, the FOS has the potential for volumes of itself to behave at least somewhat like semi-solid bodies or objects. Transverse waves, like the ripples you can see on the surface of a pond, do not lead us into thinking that they can act like a semi-solid body. Or possess the properties of such bodies. While it is immediately evident that a dense region of the FOS could behave as a semi-solid body.

The form the electron takes is derived from a gamma-ray, and yet under natural (cold) conditions rarely exceeds 5% of the speed of light (the velocity associated with the emission of an electron from the nucleus[?]). At this velocity, it no longer behaves, as we are accustomed to it doing so. To behave normally, an electron's velocity in the outer regions of an atom reaches around 0.7% (approx. 2200 km/sec. Isn't this also the average velocity of galaxies?) of the velocity of light. Based on the ratio[11] of 1 over 137 - at least for the light elements. [Need more on this.] [Is this velocity of the outer electrons a limiting factor in the normal velocities of stellar matter? As this apparently is roughly double the average velocity between many bodies regardless of where they are.[12]] Also whatever its form, the structure of an electron would have to account for magnetism.

As I started to think about the interactions of semi-solid bodies what came to mind was my memory of an observation, from watching smokers, that smoke-rings (although they originate from a similar energetic puff of air) could be made to move much slower than a smoker's normal exhale of smoke. [Smoke ring image to come.] The reason being that the volume of smoke is expending a fair amount of its energy in flowing around itself. I speculated that given the fluidic nature of the FOS, that such a form could be brought into existence within the FOS by the collision between two, different enough in terms of wavelength, gamma-rays (see A thru G). Their size would have to differ enough so that one was more solid than the other, and to give the denser of the two something (volume-wise) to work with. Well, that is at least how I envisioned the preconditions. As a result of the collision, a remarkable transformation takes place that alters their structure, and yet changes their characteristics remarkably.

---

[11] see a textbook for a more detailed discussion on this subject

[12] See The Big Bang Never Happened section Too Big For The Big Bang

No explanation for what these changes could be has ever been given until now. The following representation of two gamma-rays is from an old thought experiment of mine. First let us remind ourselves that today the data indicates that photons are simply a longitudinal wave that triggers a higher-than-normal density-pressure region (compression) followed by a lower-than-normal density-pressure region (rarefaction). Tracking a photon's effect on the aether while passing through it generates the typical representation of an electromagnetic wave by a trigonometric sine waveform of initial maximum compression followed by maximum rarefaction.

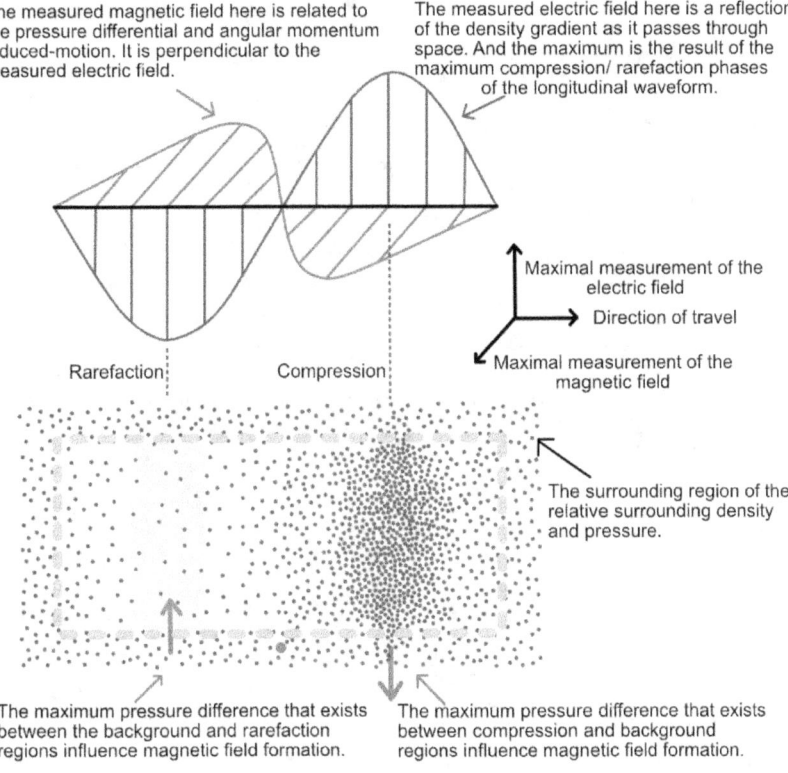

Figure 30- A transverse wave form mapped from the components of a longitudinal wave

Two gamma-ray physics is not about how they give rise to an electron-positron pair - the mechanics, but instead up until now has only been an observed fact. Only by using a longitudinal density waveform can a mechanical explanation even be tried. A transverse boundary waveform does not provide a mass to work with to form anything.

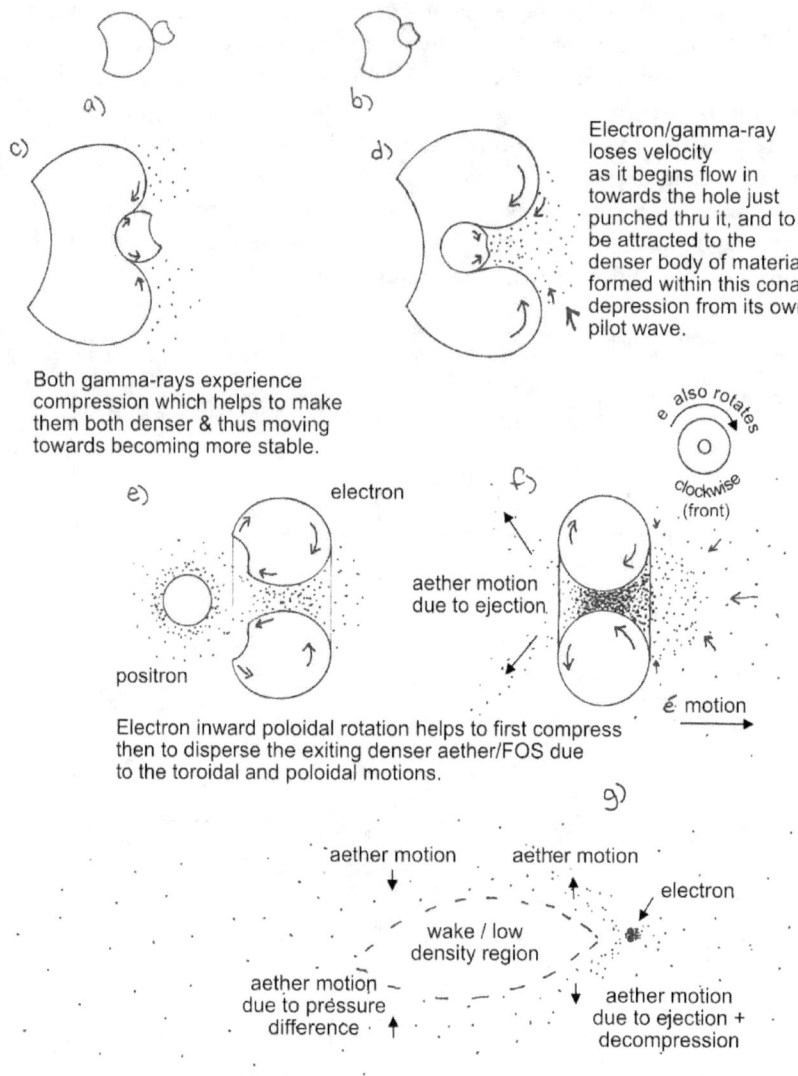

Figure 31- Two-photon physics giving rise to an electron-positron pair

Figure 31 shows the collision of two dissimilar sized gamma-rays where the smaller, and thus denser one, penetrates and passes through the larger one. A smoke ring-like body is generated out of the larger body, and a standing-wave is formed by the smaller one. Giving rise to an electron and a positron.

In the beginning, the collision proceeds as any such encounter would between two dissimilarly dense and unequal sized bodies. With the smaller, denser of the two, initiating penetration of the larger, less dense body (see A).

As the larger photon is penetrated, its outer forward surface is drawn into itself by "following" the smaller but more energetic gamma-ray (see B). The penetrating gamma-ray is being transformed into a positron, as it passes through this body of the forming electron, by being dramatically compressed in the process. Apparently resulting in it becoming a standing wave in the fabric of space. Could it be that as it passed through the electron some of its own wave energy was diverted from directly ahead of its path of conduction to the surrounding body of the gamma-ray it has just passed through? Internally changing the path of conduction from directly ahead of itself to its exterior sides of its outer self. After the passage of this newly formed positron, the wave energy of the newly formed electron is now conducted back to the only remaining body of dense FOS around. That is itself. The photon's velocity, the energy of propagation, being translated into circular motion within & around itself. This is triggered by the poloidal and toroidal rotations of this electron's body. However, this inward rotating "smoke-ring" is not entirely totally self-attracting. The wave energy traveling now back to the front of the electron is also partially attracted to the FOS ahead of the electron as is any wave attracted to denser material ahead of it, but then once it reaches the front, it finds itself attracted back into the cone-like depression or the hole of this torus. Drawn within because of the induced internal rotation, but also because here the fabric of space is accumulating and becoming denser. Due in part to the necessary compression of the fabric that is building up ahead of it in order for it to pass through the center of this electron-torus. This compression gives rise to a new region of FOS in front of the electron for the electron to be partially conducted towards. This forming gradient begins to be pushed ahead of the electron, in turn, initiating compression of the fabric ahead of itself. Like a "solid" body moving and compressing the air ahead of itself. It is brought about by the resistance of the fabric to being set into motion (its inertial property). All of the relevant fabric of the forming "gradient" tries to move in with the surface responsible for its present state of compression.

The existence of this compressed material ahead of, and within, the electron has the effect of producing a new region for influencing the conduction of the photon-electron. The focused direction of travel is no longer infinitely ahead of the photon-electron, direction wise; instead, it is just a short distance away at this moment for now on. The fabric that passes through this core is at first compressed then as it exits the electron it finds that there is no longer a mechanism of compression upon it and thus the fabric starts expanding back to something akin to the surrounding region's normal density. But due to the rapid velocity of the fabric's expansion, and the "hole" (or absence of the normal fabric density) left by the passage of the electron through the fabric of space, it does not simply stop once it has reached this trail's potential normal density and thus over expands. [The electron's spin it inherited from its parent photon also imparts angular momentum up the exiting material.] Producing a wake immediately behind the electron of a lower density and physically more energetic, active nature. The fabric exiting the core of this torus-electron, is expanding behind the electron and so is of a lower density nature. And thus, around atoms, and within nuclei, the electrons disrupt any density gradient, and therefore these regions are nowhere nearly as conducive to the conduction of electrons as other regions, they have passed through less recently and thus these other regions are more likely to have their gradient more restored and be more effective at conduction. Thus, more likely to draw electrons in them. [Could introduce the various orbitals and quantum regions here.]

This model of an electron gives us a photonic body that is automatically conducted towards the denser regions of space that it encounters, but can be disrupted from its potential regular path by interacting with both increasing and decreasing density gradients, photons, a near collision with other electrons and magnetic fields. [Should we introduce, in this section, the fact that quantum fields/regions are just density variations that increase or decrease the statistical likelihood of finding an electron in these regions?] The wake dissipation regions that form become the boundaries between the regions where electrons are more likely to be found, and thereby give a mechanical solution to Erwin Schrödinger's electron wave functions, whose intensity at any point indicate the likelihood of finding an electron there.

A second method of producing an electron that has been observed comes about when a gamma-ray travels into the domain of a nucleus. With the photon passing through the oncoming, rapidly revolving (on the order of 1/2 of 1022 revolutions per second) and the densest portion of the FOS density gradient around a proton, or group of protons. A gradient of the aether fabric forming around nuclei of a similar density as that of the gamma-rays. The evidence and empirical data arising from the photodisintegration, aka photofission and photonuclear, experiments on various nuclei by the bombardment of nuclei by various gamma-ray sources. This type of experimentation is used to see what gamma-ray energy is needed to dislodge protons, neutrons, alpha particles, and others from nuclei. [See the JENDL project and IAEA for more information.] Thus, such a gamma-ray encounter, with the right nucleus, giving rise to the collision between similar bodies of FOS, bringing about the production of an electron-positron pair. The gamma-ray experiments, are like the photoelectric effect experiments, are a way to measure what strength of photons, in this case, gamma-rays, does it take to eject particles out of the nuclei of the various chemical elements and some of their less common isotopes.

It is the electron's drastic reduction in velocity through increased self-attraction that makes it extremely sensitive to even minor FOS density gradient changes and de Broglie's [& Deutch's] pilot waves. In part because these electrons have so much more time to be conducted by FOS density variations. The resulting more easily influenced path of conduction and thus stronger attraction for any greater FOS density variation of space ahead of it, or repulsion away from lower density regions, in conjunction with the electron's over expanded lower density wake region, allows electrons to be the effectors of "gravitational" attraction between all bodies of atomic matter. These actions also allow electrons to bring protons together to form nuclei, atoms and molecules. In other words, this attraction and wake action is also responsible for the strong nuclear force and the electro-weak force. The latter giving us chemistry. What varies between these all is merely the degree or intensity of the activity of the electron's disruptive pressure reducing wake and its' favoring one side of an atom or its closeness to protons within nuclei. The gravitational effect, in the end, is merely an electron distribution function around atoms triggered by their being in a FOS or aether density gradient.[Classical electron radius equation?]

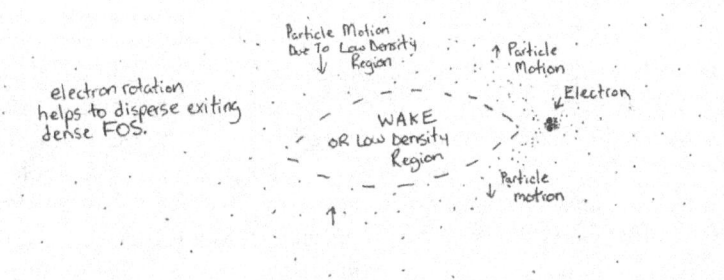

Figure 32- Electron generates a wake dissipation region as it passes through space

How electrons do this is by spending more time on the side of atoms/nuclei that are closer to another gradient source or even just due to the gradient reforming more rapidly on the side of an atom that is closer to a larger body's gradient/gravitational field. Consider the fact that in lightning events, we can detect that the negative charges move to the bottom of the clouds, while the positive charges move to the top.

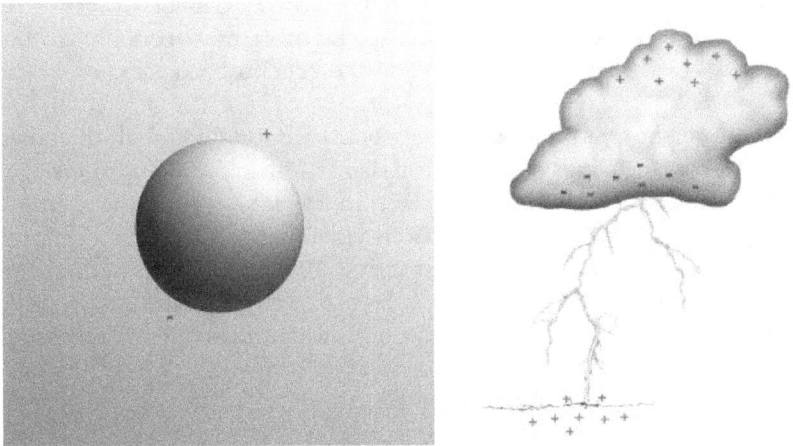

Figure 33- Atomic and cloud charge distribution within the Earth's gravitational influence

Keep in mind the difference in strength between the electric force and the gravitational force. The difference in magnitude is on the order of 10^39 times. So, what percentage of electrons favoring one side of an atom towards a gravitational mass would be required to have two 1.0 kilogram masses a metre apart move towards each other? [More to come.]

Some of what I am talking about can be seen in the electron probability distributions. [Need to expand upon this. Which might also be a way to find evidence of what I'm talking about or trying to describe.]

Someone at this time might try to point out that electrons do not enter nuclei so how could they possibly be responsible for the strong force let alone the electro-weak force. [Electromagnetism combined with the Weak force for radioactivity.] There is overwhelming experimental and observational evidence that shows that this is not only a false statement, but it has been known to be so for a few decades now. It seems to be simply ignored. Why? Is it another inconvenient fact? The source of this evidence is a form of radioactivity called Electron Capture. Formerly Electron Orbital Capture. There are a number of neutron deficient isotopes for most of the elements in the periodic table that will absorb an electron that reduces the proton count by one and increases the neutron count by one. Pure electron absorption is the least energetic outcome for the least neutron deficient radioactive nuclei, while the most neutron deficient nuclei can trigger other forms of radioactivity, including positron emission.

How common is this form of radioactivity? If you get a copy of the CRC Handbook of Chemistry and Physics and look at just the first 54 elements of the periodic table, up to and including Xenon, there are 251 isotopes in which Electron Capture can be a form of radioactivity. "Can be" because some isotopes can also decay by other means. The other most common forms of radioactivity being beta electron emission and beta positron emission.

Other's might try to refer to the Classical Electron Radius which shows by calculation that the radius of an electron is greater than the radius of the lightest of nuclei like deuterium, tritium, He-3,… to try and convince themselves and others that in no way can such electrons enter and reside within such nuclei. But here again, they would be wrong because the equation relies upon the mass of an electron in the denominator of the equation. The fatal flaw being that muons are considered just energetic electrons with a mass of ~207 times greater than that of an electron. And pions decay into muons. The rest mass of a negative pion is ~273 times that of an electron.

$$r_e = \frac{1}{4\pi\varepsilon_0} \frac{e^2}{m_e c^2} = 2.8179403227(19) \times 10^{-15} m$$

[From Wikipedia for the Classical Electron Radius.]

Classical Electron Radius: $r_e = 2.8179403227(19) \times 10^{-15}$m
Proton charge Radius: $r_p = 0.8751(61) \times 10^{-15}$m
The electron thus has a radial difference ~ 3.2199 or 3.22 times larger than a proton.
Using instead the masses of the more energetic forms of electrons:
Muon rest mass yields: $r_{mu} = 1.3589 \times 10^{-17}$m
Radial difference ~ 64.402 or 64 times smaller than a proton
Pion rest mass yields: $r_{pi} = 1.0305 \times 10^{-17}$m
Radial difference ~ 84.926 or 85 times smaller than a proton

[Proton radius compared to electrons, muons, pions]

Proton	Electron	Muon	Pion
0.8751×10^{-15}m	2.8179×10^{-15}m	1.3589×10^{-17}m	1.0305×10^{-17}m
	~3.22 x larger	~64 x smaller	~85 x smaller

Table 5 - Proton radius compared to electrons, muons, pions

Now someone might want to point out that muons have never been observed entering nuclei in experiments that have been set up to test this and thus cannot be the source of the strong force. Actually, if you break matter in experiments with antimatter, it is negative, positive, and neutral pions that you end up with. Even so back to the muon argument – it comes back to the failure to realize that it is only neutron deficient nuclei which can accept/force electrons into them. That is the essence of the form of radioactivity called Electron Capture. As all other balanced and thus stable nuclei possess a nuclear potential barrier that is in effect like the full electronic shell of any atom. If it is occupied, and not of the correct density-gradient nature, then such a nucleus will not accept another electron, regardless of its energetic state of being either a muon or a pion.

So, electrons can enter nuclei and in the energetic state called a pion they can inhabit nuclei and are considerably smaller than protons. If the Classical Electron Radius equation is valid within the nucleus, which is doubtful, then this equation by which those who refute the ability of electrons to enter nuclei, may still prove useful as a source of evidence about the true physical nature of nuclei and electrons.

Velocity is known to increase the mass of an electron; then perhaps we can use the apparent increase in the mass of an electron within the nucleus to determine the speed of them within nuclei?

Neutrinos are intimately linked to electrons and something which is considered astounding is that anti-neutrinos from the hypothetical weak force reaction triggering beta decay, of nuclei like Cobalt 60 into Nickel 60, violate the mirror symmetry assumption or principle. Neutrinos, it has not been fully determined* yet if the anti-neutrino is different from regular neutrinos, generated in nuclear beta-decay were found to have what is called a left-handed symmetry at all times which turned out to be not only a complete surprise but also violates the assumption of parity symmetry. The idea that parity conservation might not hold first arose when theoretical physicists Tsung-Dao Lee and Chen-Ning Yang reviewed the work of other theorists on the question of parity-conservation in all fundamental interactions in nature. They concluded that in the case of the weak force interaction, that the experimental data neither confirmed nor refuted parity-conservation. They then asked experimental physicist, Chien-Shiung Wu, for her help in investigating this. She was an expert on beta decay spectroscopy. Professor Wu, with help from Henry Boorse and Mark W. Zemansky, setup an experiment to test mirror symmetry and shockingly discovered that all neutrinos in beta decay have left-handed spin. It was at first not believed by others in the field and was repeated by other experimenters and the results held. [Add more on the Wu and related experiments.]

In the FOS [fabric-of-space] model introduced in this book this left-handedness of neutrinos is complimentary with the model. More evidence of this model's validity. The behavior of electrons within atomic magnets and electro-magnets display a spin complimentary to the left-handed spin, counter-clockwise rotation, about their axis of travel. Beta-decay of nuclei shows that these events allow the electrons to eject a small quantity of aether that reflects this left-handed spin as it leaves the nucleus and is no longer able to support the density of the material it had been working within in the nucleus. [Is this the place to introduce what a neutrino is? That is the data seems to imply that neutrinos are simply a type of volumetrically small photon?]

In another chapter, we will look at how this "smoke-ring" structure, in relation to electron spin, helps to produce the force known as magnetism.

If electrons are the effectors of "gravitational" attraction then what role do protons play? And how do electrons & protons interact with one another? How do we get from positrons to protons?

The Proton/Positron and Mass

We've theorized that the nature of positrons is due to their origin as standing waves. And this leads to them eventually forming protons, which by mass, is the main constituent of any gravitational body and thus the main source of a "gravitational field". What is the difference between a positron and a proton? If you break a proton, you end up with a positron and the emission of gamma-rays. Somehow the presence of the gamma-rays is indicative of the greater mass of protons. And the wavelength of the emitted gamma-ray a measure of its density. This increase in density makes sense if the standing waves are beginning to accumulate material around them. The Standard Model would have us believe that the positron is much less than a proton, but the Standard Model cannot account for why in the destruction of a proton that just a positron and gamma-rays are released. And more importantly, it cannot answer the question - What is mass? The standard model would have us believe that through the Higgs field and Higgs bosons that particles gain mass. But it still does not tell us what mass is. Interestingly enough, the Higgs model proposes adherence and crowding around a particle to produce a mass effect, which is similar to the FOS model and aether compaction around protons, or ahead of electrons. And through the FOS model series, we also get Einstein's relativistic velocity mass increase at the same time due to a resistive aether filling all of space.

In considering a longitudinal density wave structure for photons, the only form a proton could consist of was a dense body of FOS, which had reached a critical self-sustaining standing-wave nature into permanent existence. This standing-wave high-density body appears to succeed in increasing the density of the fabric around itself for quite some distance. The evidence begins with the photoelectric effect. As for more closer to the nucleus if you use photonuclear data for deuterium [heavy hydrogen], or the nuclear binding energy, that density appears to be roughly the same density of the gamma-rays that can trigger an electron-positron pair event or break apart deuterium. The nuclear binding energy of 2.2 MeV as opposed to the first outer electron with an ionization energy of 13.602 eV. [Versus regular hydrogen's 13.598 eV.] In other words, the photon energy required to eject a nuclear electron is the same. The photoelectric effect for nuclei. [More.] Why would the fabric become denser in the first place? And yet it would appear to do so, for this is one of the key ideas of the photoelectric effect of the FOS model, and in conjunction with known fluid dynamic behaviors I was able to expand the theory to cover most of the observed phenomenon of the universe. Only recently, at basically the end of the development of the model (for my part), did I come up with a possible explanation as to why the density of the fabric would increase around the proton at its formation, and continue to maintain that effect upon the surrounding fabric. The hypothesis that seems to account for the gradient effect, around a proton and positrons, is based upon the physical dynamics of the surface of these standing waves and their spin.

A proton, or positron, is formed when a volume of FOS becomes compressed roughly into a spherical undulating body. Any wave energy brought into the proton at its formation remains within it like an elastic ball that experiences no losses as it "bounces." But this ball is floating in outer space. A standing wave is formed. The density of the internal fabric confines these waves internally by being locally the only volume of material attractive (greatest potential conductor) to these waves. Consider that a wave in one medium is reluctant to leave it when the surrounding medium is not as conductive to that wave or too great a change in density. Like the ripples in a pool where the energy of the waves has a better mechanism of propagation within the pool and less able to transfer the energy to the air above it. And so, the wave energy is confined to the liquid of the pool. [Not the best analogy given that it refers to transverse waves.]

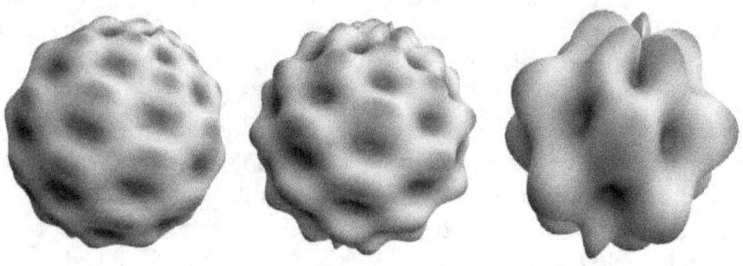

Figure 34- Standing wave points in space

Figure 34 shows simulated standing wave bodies using Mathematica by Wolfram, but they do not show any spin related components. Compression of the surrounding material around them being triggered by the varying pressure waves being generated by the surfaces of these undulating standing waves. With the core being effectively more uniform in density and the outer layer varying in density. Thus, this is likely why they are not seen as solid objects, and their boundary layer appears as more solid objects fluctuating in and out of existence. Appearing momentarily as solid bodies.

So now consider a volume of material which is totally self-absorbed - no wave energy escaping. A volume in which perhaps a wave never gains enough momentum/acceleration to carry itself beyond the borders of the positron core of a proton. Too great a volume during formation though and too much of a wave forms, tearing the volume of fabric apart (limiting the size of positrons & protons). Are positrons formed from specific wavelengths more likely to become protons than others? Does it matter? Once again if we look at radioactivity, the answer would seem to be - No. The evidence coming from positive beta particle emission out of nuclei who formerly were too neutron deficient/poor. These nuclei can stabilize either through Electron Capture or positive beta particle emission. It would seem to be that some near-collision by two protons, within the nucleus, triggers two gamma-ray dense bodies to interact within a nucleus and generate an electron-positron pair in the process. This is the only reasonable possibility as outside the nucleus two-photon physics does exactly this.

Regardless of where a positron is formed it seems we end up with a roughly spherical body, whose spinning in conjunction with its undulating surface, sets up a mechanism by which the positrons can induce the formation of secondary waves, similar to pilot waves, in the surrounding medium - the fabric of space – and trigger the formation of a density gradient within the aether. These secondary waves are probably distorted by the positron's spin, and might be the cause of the magnetic moment of a proton.

Consider the possibility that the fabric [aether] density gradient surrounding the proton is a by-product of the secondary waves being generated by this standing wave's undulating form. In other words, a density increase of the aether surrounding a proton/positron is being triggered by the secondary waves being emanated from the vibratory surface of the proton/positron. Initially, as the waves are sent out a drop in the density of the fabric occurs since the material is locked in the formation of the denser regions, and as a result of this the outer surrounding fabric is pushed back towards the center as it tries to compensate for the lower density region which now exists between the waves being generated and the surface of the standing wave itself. In this scenario, the overall density is related only to the undulations of the protons' surface. There must be a balance between the undulations of the protons' surface, and the FOS density pressure that arises from compacting the fabric by these waves. The fabric is drawn into the region surrounding the proton, due to this process, resulting in a pressure buildup of fabric around the proton/positron.

Figure 35- Mass gradient versus an atom's quantized gradient tiered structure

Comparing a regular mass gradient versus an atomic quantum, or more accurately a quantized, gradient profile. [Image modification in progress to show electron orbitals.]

The formation of a FOS density gradient provides a physical structure for electrons to work with to give us gravitational effects, and electrostatics. But we must consider the difference between the waves generating the gradient and the accumulation or resulting density. Proton mass is related to the accumulation and formation of gamma-ray dense material around the positron, while the positive charge is related to the attempted change in pressure triggered by the waves emanating from it. Or more accurately what turns a positron into a proton is the accumulation of gamma-ray dense material around a positron. And this "pinching" of the material around it and to its current point in space is seen as mass. By "pinching" the fabric, the proton is effectively trying to adhere to a single point in space and thus gains inertial resistance. This is what the Higgs mechanism is supposed to accomplish with the Higgs particle interacting with other particles in the Higgs field. Occam's tenant would clearly choose the FOS model of standing waves which accomplishes the same thing with less complexity and explains even more. How does, or more appropriately where, does the conversion of positrons into protons happen? Inside stars? Or at the death of stars? It does not seem likely to occur anywhere else, or we would see positrons turning into protons within positron storage containment vessels. It would seem that there must be a critical "point" between just a positron and the triggering of enough gamma-ray density material around a positron to form a proton. Does the starting wavelength play a role? Why is it when some positrons undergo annihilation with an electron there are three gamma-rays produced instead of the normal two of formation? [If we sum up their masses, based on a wavelength relationship to density, do we get a similar amount of material or a different amount of material?] What role does the granularity of the FOS, based on Planck's constant, have on the charge we measure? Is there a limitation based on this granularity and reaction time of the FOS? What can we say about the fact that we have measured, or perhaps inadvertently defined, the charges to be equal and opposite? Since we see atoms that technically on these definitions should be charge neutral and thus not share or give up electrons to other atoms, for chemistry to exist, that our definition of charge is not complete. Which is why shell completion, or filling, came into existence. We now know the basis of what charge is, and now can explore in greater detail how it works.

The return to an aether, fabric of space, universe was the only obvious path to take, to explore a physical possibility for a simple model of the mechanics of the universe, and a welcome challenge for exploring the boundaries that would permit the formation of such a working model and potential theory. To begin with, this new aether model is based upon a fabric of space that forms a density gradient around a body. Density is the key and waves do not trigger drag. Something that is supported by the experimental detection of the "twisting/shearing of space" around the Earth by Ignazio Ciufolini [et al], Gravity Probe B and Roland De Witte. [aka frame dragging] Possibly something to do with the Earth's own pilot wave forming on one side and triggering a gradient difference that has the influence of appearing as twisting.

Mass is then a particle's adherence to the fabric it is within. And thus, we now know why an elevator's acceleration is hard to determine whether it's due to its position on a planet or its acceleration in free space. But an aether is required for this adherence for inertia to exist. If we truly had empty space, then mass should not increase as we approach the speed of light while alternately it makes sense if there is an aether to resist an increase in velocity by mass increasing due to the fabric's resistance as a particles' velocity approaches the speed of light. No such explanation can be given under the current accepted model of a non-empty empty space – a contradiction by definition. The most popular theory invokes a new particle and field to give us mass. The Higgs Boson and the Higgs Field. And once again with the Higgs mechanism, we have something that still cannot produce a physical explanation for why mass and inertia do what they do. What happened to Einstein's curvature of space being the source of the mass or at least an indicator of its presence? His curvature equation is simply the equation of the gradient around a mass.

From Wikipedia 20161204

"The Higgs field is an energy field that is thought to exist everywhere in the universe. The field is accompanied by a fundamental particle called the Higgs boson, which the field uses to continuously interact with other particles. As particles pass through the field they are "given" mass, much as an object passing through treacle (or molasses) will become slower because they now cannot travel at the speed of light because they have mass."

Gravity is just the outcome of electrons very weakly favoring one side of an atom. In the purest sense "mass" does not bring bodies together, mass attracts electrons, and the electrons bring together masses.

Gravity, Charge & Protons

When a pigeon picks up a piece of bread to eat, that someone has thrown at it, it has successfully pulled that bread away from the gravitational pull being generated by our planet. It has "defeated" the Earth despite the Earth's entire mass "pulling down" on that piece of bread. That is how weak gravity is. Remember that compared to the electric force that binds electrons to their parent atoms that gravity is weaker by a factor of 10^{37} times [?] smaller than the strength of electrical "force." [Missing Newton's action at a distance.]

Sir Isaac Newton's law of gravitation forced Newton to utter the famous quote: "Hypotheses non fingo." Latin for basically "I feign* no hypotheses", just as Johannes Kepler did, in regards to how gravity worked.

From Newton in 1726:*[Wikipedia sources quote.]

"I have not as yet been able to discover the reason for these properties of gravity from phenomena, and I do not feign hypotheses. For whatever is not deduced from the phenomena must be called a hypothesis; and hypotheses, whether metaphysical or physical, or based on occult qualities, or mechanical, have no place in experimental philosophy. In this philosophy, particular propositions are inferred from the phenomena, and afterwards rendered general by induction."

The Death of the Dark Energy Idea

To Newton and Kepler, the universe seemed to obey the laws of mechanics, and these laws could be observed in the dynamics of matter in motion. But how did gravity work between distant objects without some evident mechanical mechanism between them? Many people hypothesized that the forces had to be mediated by an aether. "Any cause and effect without a discernible contact, or action at a distance, contradicts common sense and has been an unacceptable notion since antiquity. EB" This action-at-a-distance, not to be confused with spooky action at a distance and quantum entanglement, was a real problem for theorists and until field theory came along from Maxwell, and Einstein, it seemed to do away with the required aether as a mechanism for it to somehow give us the forces. Not that any mainstream theory yet explains any of the forces and instead simply puts off an explanation by saying that the "force" is being emanated or generated by a particle, an inter-dimensional string or some other speculative theoretical construct. This is the problem of most of contemporary physics. Experimentation has allowed the development of equations that allows the user to make predictions, but no one actually ever gives an explanation for how any of the forces actually work. Einstein's field theory of a distortion in time and space did away with Newton's action-at-a-distance conundrum but then where Einstein's model failed as other explanations before his in explaining how gravity actually works has now been, or so we are told to believe, solved by the Higgs mechanism acting on matter through the Higgs boson within a Higgs field. There is no curved space in the Higgs explanation. Nor does it predict gravitational waves. Gravity is no more clearly explained than it was in Kepler's and Galileo's time of the 1600s. The whole Higgs explanation is no better, but ironically points towards one aspect of a similar explanation to adherence to a point in space like the FOS theory series described here within these pages. Whereas with Einstein's explanation, utilizing the gravitational constant to formulate a gradient field, we at least have more accurate equations compared to Newton's, and his also predicts the growing inertial resistance of matter as it approaches the speed of light. [Was the preceding statement accurate in its choice of words?] Remember though that Newton's equations are still accurate under most circumstances including putting a satellite into orbit around Saturn,...

To explore gravity and the interaction between electrons and protons, we must first return to the structure of a proton that the new model is proposing. For it is here that the true nature of one half of the cause of the electric force termed "electrical charge" originates, and it is the protons who are also the main source of the mass that is responsible for gravity. What is gravity? Gravity is by definition, the force that exists between electrically neutral masses that brings together those masses. It is the force that pulls down apples from trees and keeps the Moon in orbit around the Earth. What is mass? I'm not talking about - "What does it do?" We know what the consequence of accumulating large quantities of mass has on a region of space and the influence it has on any electrically neutral objects nearby. Mass is also what electrons are attracted to and thereby draw in any body under their influence towards the center of mass. In this new model, we have defined the proton mass as the consequence of its wave energy triggering compression around itself effectively "pinching" itself to the point in space it occupies and/or continue to conduct itself along the current path it finds itself traveling [due to the density wave it is triggering] and thus maintaining its constant velocity.

The electron is attracted to the FOS density gradient of the proton, which to it is one outcome of the positive aspect of the force/property known as electric charge. *Remember that a positron has the same charge as a proton, but its mass is different.* The structural aspect of the proton, and positron, that produces the effect, known as charge, is the rate of change in pressure upon the surrounding fabric (by the compaction of it) as it moves in to fill the spaces between the emitted waves produced by the proton's undulating (resonating) surface and rotation. While the mass of the proton is the resulting building up, or pinch, of gamma-ray dense material around it. Resulting in the formation of an atomic FOS density and pressure gradient. Remember if you break a proton, often done in anti-matter/matter annihilation experiments, you get a positron and some gamma-rays. So, mass is related to and indicated by the presence of gamma-ray dense material. [Could the formation of a proton versus a positron be a function of the starting gamma-ray wavelengths and perhaps their start/formation inside of the gradient density fields of stars?]

When an electron passes through space, it leaves behind it a dissipating wake-region of a lower density and lower pressure FOS. Which if close enough to a proton provides a volume of space to which the proton, under the right conditions, can be pushed towards this lower density region by its own FOS density gradient pressure on the side of the proton that is opposite it in relation to the transient wake left behind by the electron. Analogous to a high-pressure volume of gas expanding and thus pushing some object in its path into the lower density volume of space. This is why protons are attracted/forced into such regions. These wakes, FOS gradient pressure reducers, are the negative aspect of the force known as charge for which the electrons are responsible.

This balance between electrons and protons is tempered by the mass-density that a proton triggers around itself, and the rate of its reformation – positive charge. The mass of a proton seems to be limited to the wave density nature of the proton's core. If the mass increases then the standing wave energy of the proton can tear itself apart. Thus, no matter can exist, except for the briefest periods of time that is greater than the density or size of protons. This is why most of the massive particles of the standard model are not stable outside the nucleus and simply tear themselves apart after their formation within particle accelerators and after any high energy collisions, and this implies no black holes. Can these particles be said to exist at all when in reality they're just temporary high-density nodes within the fabric of space that tear themselves apart due to the instability of their high-density nature and size.

A typical hydrogen atom is formed when an electron enters into orbit around a proton at approximately of a distance of 0.35 x 10^-12 metres. [Not Bohr's radius. Bohr's radius was based upon so-called fundamental properties applied to a well-understood equation. It fails to accurately reflect reality when it actually comes to the true radius of the first likely location of an electron around a hydrogen atom. It predicts the radius to be of 0.53 x 10^-12 metres.] The electron takes up orbit around a proton because it is attracted to the [forming] FOS gradient of the proton (the positive aspect of the force of charge). And moves not much closer in, or farther away, because at this distance its conduction towards the proton is restricted by its own momentum. In hydrogen alone, momentum would seem to be the only control preventing an electron from shell 'K' from moving in to become part of the nucleus. In the volume of material, it inhabits at a density at 13.598 eV at its most probable orbital distance from a proton, the electron's conduction strength cannot be changed by the rate of change of the gradient located here. Not until we get to Beryllium-7, $^{7}_{4}Be$, as opposed to the normal form of Beryllium-9, $^{9}_{4}Be$, the fourth element on the periodic table that this form of Beryllium, which is neutron deficient by a count of two, that an electron from a 'K' shell can have its conduction energy strength change and overcome its momentum to enter into a nucleus. Converting the atom in the process into normal Lithium-7, $^{7}_{3}Li$. In other orbitals [L,M,N...] what prevents electrons from traveling from one orbital position, into an orbital closer to the nucleus, is the wake dissipation region of the shell below it (closer to the nucleus). The lower density and disrupted nature of this region, triggered by the electron closer to the nucleus, deters the conduction of an electron into a shell closer to the nucleus. This appears to be because each electron's conduction strength is proportional to the region it is contained within and in part due to the wake region below it. Thus normally to an electron, the potential orbital below is not as conductive to it as the region it already resides within. [Which is why we end up with quantized orbitals and quantum mechanics.] But if an electron were ejected, or lost, from the shell below, the wake dissipation region would become more conductive, and thus permit an electron to pass through the former wake region, which has now become a better conductor, and allow the electron to take up residence. Which in turn would then return the wake dissipation region, it had just been conducted through, back to the formidable barrier to other electrons that it once was. Doing so by adding its wake energy to the reformation of the wake dissipation region that forms above any nuclear encircling electrons. The structure of the surrounding FOS gradient actually becomes of a stepped, or terraced structure of levels of higher FOS density regions separated by lower density wake regions (see the next figure).

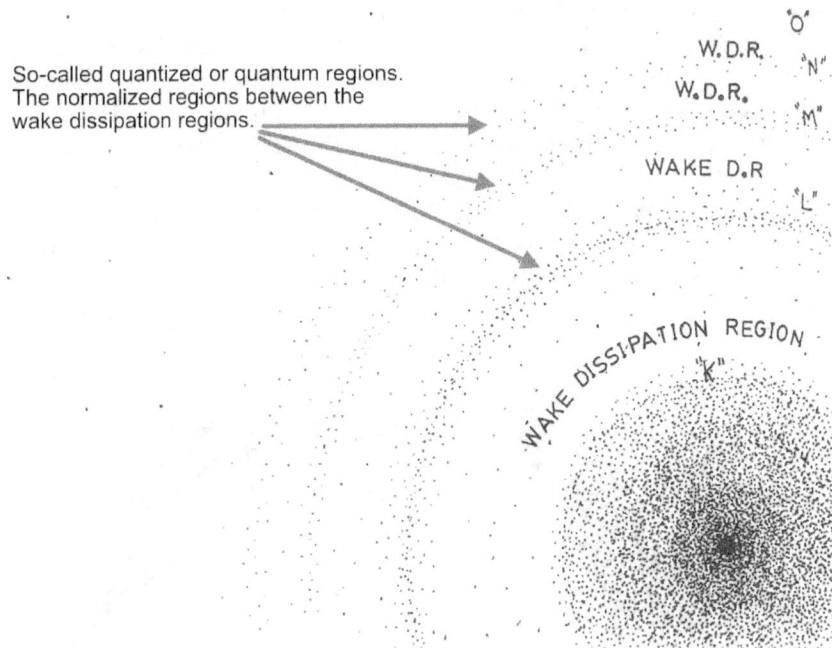

Figure 36- An atom's electron generated wake dissipation regions [exaggerated]

Within the shells, the individual electron orbitals remain separated, and terraced as well, for similar reasons, but the wakes have not yet expanded enough and combined together to form a larger scale bar+rier. Like that separating the main shells (K, L, M, etc.). So electrons can orbit around one another fairly close to each other but do remain in separate orbitals. A lone electron's wake seems to be just active enough to lower the density of the fabric of its own orbital such that it shall have just returned almost back to its normal density by the time the electron has returned to this region since its last orbital revolution. But not quite. The result is that the electron will find paths somewhat parallel to its last one more conductive thus forcing it into "precession" about the atom. This forced precession can in some cases be overcome by placing the atom within a magnetic, or electric, field to counter the electron's attraction elsewhere. In magnetism, this precession can only be overcome if the surrounding FOS gradient is not great enough to have forced the two closest electrons into being what is termed a "pair". So-called pairs of electrons are too close together such that the upper electron's own wake also reduces the FOS gradient characteristics of the electron's domain below its own. Because the material is expanding to fill in the wake region of the electron above - this action further reduces the regions gradient. Forcing further conductive induced precession as the electron, closer to the nucleus, is drawn to regions that have had greater time to recover from its own wake as well as that of its neighbor's wake above. If the spacing of the electrons is greater though there is less expansion of the FOS within the lower shell induced by the passage of a wake from above. Due to the fact that there is that much of a greater amount of the fabric between orbitals to expand & re-normalize the gradient. Since there is less expansion of the lower electron's own gradient, this must nearly eliminate any induced precession that may have arisen from this effect.

    Any electron above cannot move in closer, because it is better suited for the density of the material of the orbital it is within. And is not close enough to being of the same energetic structure as the electron below, partly for the reason that this lower electron's wake is lowering the above electron's FOS gradient region as well. Thus, keeping the upper electron as it is - energetically speaking.

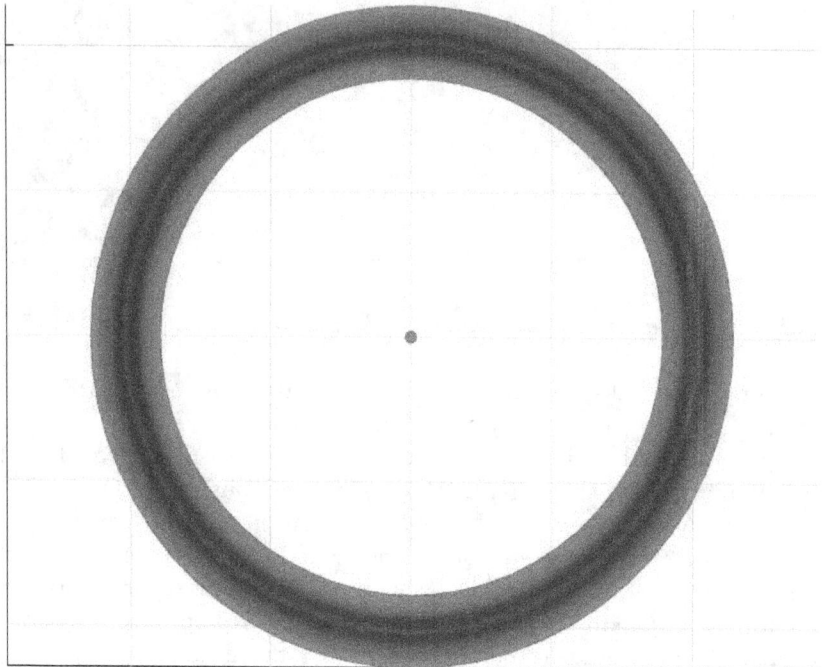

Figure 37-probability map of the location of a particle from a simulation using GNU Octave where its path is fluctuating due to particles affect from its last orbit

In the figure above, generated using the GNU Octave programming language and interface, we are shown the probability of finding a particle whose changing orbital radius is due to its particles residual effect on the region of space it had previously passed through and is continuously returning to. Since the aether gradient has not fully recovered from the effects of the particle's previous orbital passage through this space, the orbital path and radius fluctuate. In this 2D simulation, the radial fluctuation is about 12.5% and triggered by a conduction fluctuation between 60% and 100% that of normal. It depends on how long ago a particle had passed through it and its degree of normalization.

Thus, we have a mechanical mechanism for quantum orbital probabilities based on physical changes to the aether gradient. No magic is required.

It would appear that the electrons contained within a nucleus also generate a wake dissipation region about the nucleus. [Not only as an important part of the nuclear potential barrier but also in keeping the nucleus together.] At first, this did not arise out of the data when considering the hydrogen nucleus, which normally contains no neutrons and thus could not have any nuclear electrons. And thus the question arose: Why are electrons not drawn into this hydrogen nucleus? Simple- the fabric is not dense enough to attract a normal electron any closer about a proton and hold on to it. In other words, the density is not great enough to affect a structural change in the electron to make it more energetic to draw it more deeply into the FOS gradient. The momentum and the kinetic energy is a factor here, but only because the structure of the electron does not permit it to possess a greater attraction towards the nucleus in its current state. An increase in the strength of the FOS gradient must be of a value that is great enough to force a structural related conduction change in an electron to overcome its momentum to move in any closer. A change that if possible, for some isotopes for most elements, will take some time and this time will vary depending on the isotope. Beryllium-7 is the first such nucleus with a half-life of about 53.28 days. While the next possible isotope that might decay by Electron Capture (EC) is Carbon-11 with a half-life of only 20.3 minutes. There is a lot of empirical data for this form of radioactivity that is called Electron Capture – EC. [Formerly called Electron Orbital Capture or EOC.] How common is this form of radioactivity? If you get a copy of the CRC Handbook of Chemistry and Physics and look at just the first 54 elements of the periodic table, up to and including Xenon, there are 251 isotopes in which Electron Capture can be a form of radioactivity. "Can be" because some isotopes can also decay by other means. The other most common forms of radioactivity being beta electron emission and beta positron emission.]

Electron Capture flies in the face of many scientists, and educators, statement that electrons cannot enter the nucleus. It just does not happen, according to many of them. Yet it is one of the more common forms of radioactivity. Ordinarily, an electron is unable to enter into a nucleus. But any nucleus large enough whose fabric is too dense (because it has too many protons versus "neutrons") will trigger a structural change in a K-shell electron to draw it in to correct this "condition." It dawns on me that this form of radioactivity may also be proof, or at least a chink in the armor of, against the "perfect" pions. An electron is being drawn into the nucleus to stabilize the nucleus by turning a "proton" into a neutron in Electron Capture (EC). If the strong nuclear force is exerted by a nuclei's pions then where is the pion to come from if it is not itself a function of the electrons of the neutrons that have been "added" into the nuclei who possess too many protons. An electron is absorbed, and the proton count goes down, and the neutron count goes up. At least for the lower energy EC events. If the energy of the event is great enough due to a greater neutron deficiency, then after the EC event sometimes positive beta particles, positrons, can be emitted. But is there any evidence for a wake around neutrons? Yes. See the following based on an image from the book Nuclear Structure to Cosmology. More ammo for the FOS theory series concept of nuclear electrons.

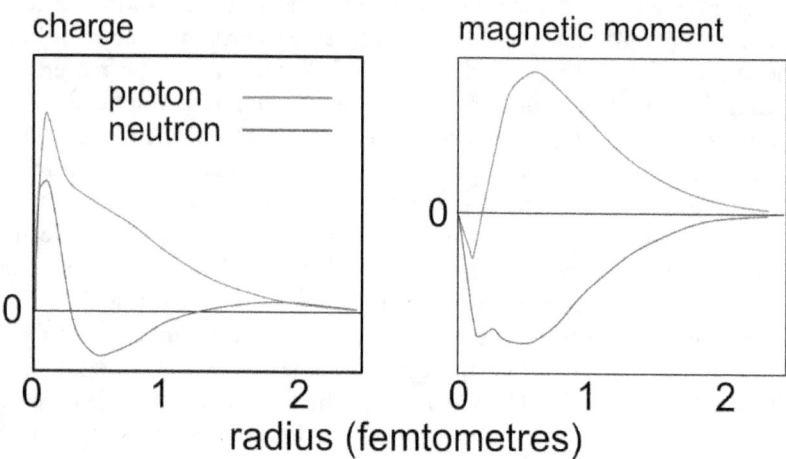

Figure 38- *The derived charge and magnetic moment densities of the proton and neutron*

Muons (heavy electrons, but not nuclear pions), which are not considered mesons, will not normally enter a nucleus because in all likelihood they too are not able to move in closer to a nucleus unless it is neutron deficient. Any electrons, regardless of their energy, orbiting a nucleus would also encounter the nuclear wake dissipation region (the Nuclear Potential Barrier) that is a product of the activity of the nuclear electrons (pions). Because the conductive nature of these muon-electrons is more pronounced, they are better at being conducted closer to nuclei and about such regions, before normally decaying, because they are more suited to the greater density of the fabric of space found here. These muons will decay, lose energy, outside of the appropriate conductive region as they must inhabit a fabric of space approximately equal to the same density in which they were inadvertently developed for. This energetic electron must reduce its conduction energy to reflect the FOS that it is presently inhabiting. Thus, the electron structure changes and adjusts to reflect the FOS it is within. In the process of "slowing" shedding its mass (and energy) in the form of gamma-rays. Which this material was part of the accumulated fabric at the front of any muon-electron, which the new forced velocity cannot keep all of this compressed fabric confined into the electron's frontal depression. The failure of muons to move into nuclei was considered to be the proof that energetic electrons could not be the force carrier for the strong nuclear force because of their inability of them to move into stable nuclei. We see now that the true failure was that of the theorists to consider the nuclear wake barrier, insufficiently conductive gradient and its reduction in EC radioactivity.

The conduction rate/strength of an electron is induced by the FOS density gradient of which it inhabits, which itself is brought about and based upon the size of the nucleus. The larger the nucleus, the greater the volumetric density value [amount of dense material], and the rate of change of density formation, from the compaction of the fabric from several protons in one location. The physical effect of the increasing density of the fabric upon the electrons concerned can be seen by the increasingly smaller sizes of the more massive atoms in a single period or horizontal row of the periodic table. The outer, or non-nuclear, electrons move closer in towards the nucleus as the wake regions diminish in volume as the FOS density increases, the increasing density reduces the size of the wakes. The density is increasing due to the increase in the number of protons generating more compression waves, and that is seen as an increase in the atomic mass. No Higgs mechanism required. [With the mass not proportionately getting heavier each time a proton is added due to the interference of each proton with the compression of its neighbors.] Periodically there is an increase in the diameter of a more massive atom's influence beyond the last full orbital at the beginning of each new row in the periodic table. This increase is brought about by the development of a newly formed conductive-enough region that can be utilized by electrons, and whose existence is due to the extension of the FOS gradient outwards. Extended, because the previously last wake dissipation region was unable to "eliminate" the effect on the surrounding fabric by the additional FOS gradient generating proton(s).

The number of electrons found in each electronic shell surrounding a nucleus is in general regulated by two factors; first, the number of "protons" contained within the nucleus (affecting "only" the furthest orbit (last level)), and secondly the ability of the FOS to return to the surrounding normal density after the passage of an electrons' wake. It is the speed at which the FOS can resume a relatively normal fabric density that truly regulates the number of electrons found in each shell and the fact that the number of electrons in all filled comparable levels of any kind of an atom is the same. It is here that the reason can be found for why an atom is still chemically active although by the old (pre-FOS) definition it is neutral (i.e. for each "proton" it possess an electron), and the reason why that this definition of charge neutrality is incorrect.

A shell is composed of a spherical region around a nucleus inhabited by electrons and dissipating wakes, which limits the population of the electrons. The limiting factor being relative to the area [density-volume factor] covered by the shell and the recovery time of the FOS density gradient from the wakes, which varies with the density of the FOS gradient of the relative shell. [The less dense the fabric, the longer [shorter] the recovery time of the fabric. [I am still thinking about this.] This is due to the fact that the drop in the density is less intense because the electrons of each successively greater shell have less dense material to propel, and thus the wakes become decreasingly weaker events but affect a greater volume. Just like explosions in liquids versus air versus space. [The following is a duplicate of material in next paragraph: Add to this that the path they take around their parent atom is increasing in its orbital length the further they are from the nucleus, and we get a relationship between the density and volume that gives us a predictable pattern of filling orbits.] This is why the further away from the nucleus, the shells form closer & closer to one another.

Since the area-gradient density recovery time coefficients of the relative shells are comparable, so are the number of electrons that can inhabit them. As the wake is dissipating, it appears to rise like a bubble under water. The material below expands due to the drop in pressure. Note that although a nucleus can be forced towards a wake region it is too great a mass to have its position greatly affected by a single wake before the wake dissipates, and also because the wake is coming into existence in a new area following the passage of its parent electron. As this wake rises it is producing, along with the others in their shells, a region above each shell of lower than normal density nature and/or too turbulent in terms of varying gradient structure. The recovery time of this orbital gradient material also likely induces the electron to take a slightly alternate path which is more conductive to it as it returns to the FOS still recovering from its previous passage through this region. All of this is in effect a barrier to other electrons for some distance until this boundary effect is weakened enough through the dissipation of the wakes. And above this transitional regional boundary, the FOS gradient returns to a favorable enough density and uniformity giving rise to a new potential shell for some other electron to inhabit. At least for a while. This is the foundation of chemistry.

A new shell can be formed if the intensity of the previous level of the gradient was not too close to being that of normal space. For if it is too similar to normal space, in terms of density, then the potential of the region we're concerned about is not then great enough to attract and hold an electron. But it none the less still might be able to attract an electron weakly. And on this note, it is time to discuss the structural and limiting factor(s) of this potentially active exterior (final) shell, which could belong to any atom.

The degree to which an atom is potentially chemically active, and the type of activity it is capable of (i.e. ionic, covalent, polar covalent, metallic, Van der Waal or inert) is solely based upon the outer fabric gradient surrounding an atom, which is itself based upon the structure of the nucleus.

We begin with a look at a regular hydrogen atom whose physical form is based upon a single electron orbiting a single proton at a distance of approximately 0.35×10^{-10} metres [0.035 nm?] and a strength of around 13.598 eV. [The first allowed quantum orbital position.] The proton is surrounded by an enveloping FOS gradient, which at and beyond its first shell, initiated by the presence of an electron, is still conductive enough to attract and just barely hold onto another electron at around 0.754 eV. This occurs because a single electron is incapable of lowering the gradient to the surrounding environments, normal density and uniformity. While the electron is being conducted in one region of the shell another region, in very close proximity, has been for some time dense enough to attract and hold a second electron. No single electron can neutralize the entire FOS density gradient of a proton except in the formation of a neutron where it is extremely close in proximity to it and of a more energetic and heavier nature. [But outside the nucleus without the help of another proton to generate enough of a gradient the neutron has a half-life of around 15 minutes. More on this in the next chapter.] Generally, in the formation of hydrogen, the gradient has started to regenerate because its wake has dissipated by the time the electron gets around to coming back from its last orbit. With a second electron in orbit, the gradient is more than adequately "neutralized". And due to this more than adequate "neutralization", one of the electrons can be and usually is shared with another atom - in our example another hydrogen atom. Thus, a molecule, H2, is brought about through covalent bonding. The next atom on the periodic table has its own unique FOS gradient so efficiently "neutralized" by its own complement of two electrons that it is the most inert of all the inert atoms of the Periodic Table of the Elements - the helium atom, ^4_2He.

It is the helium atom's own unique nucleus, consisting of two protons & two "neutrons" (normally) that produces this "most perfect" & efficiently neutralized FOS density gradient. But then this is not the only form & thus the only characteristics that helium can take on - for the characteristics of helium are based upon its nucleonic form (nuclear structure), and this varies with a few isotopes of helium. And as a result, of course, so does the gradient, which affects an electron's conduction, and FOS displacement characteristics, which in turn affects the chemical behavior of the atom - ever so slightly. This slight variation in chemical behavior, due to an irregular number of "neutrons" within the nucleus, is most readily observed in the heavy hydrogen atom known as deuterium. It appears to be normal until used to make heavy water, named after the change in the molecular mass of the H_20 molecule due to the use of deuterium instead of regular hydrogen, whose properties are quite a bit different than that of ordinary water (see table). This is brought about due to the fact that the FOS density gradient is slightly greater around a deuterium nucleus than a normal hydrogen nucleus (just a proton), and due to this the electrons are not as unevenly shared between the oxygen atom and the deuterium hydrogen atoms as they are in a normal water molecule. For regular hydrogen, the inner and outer electrons are potentially held with a strength of about 13.598 eV and 0.7542 eV respectively while for deuterium that strength is 13.602 eV and 0.7546 eV. It might not sound like much of a difference, but water made with deuterium is toxic to most life forms. Not like a poison but instead of a disruptor to normal chemical biological processes.

The larger the nuclear mass becomes the smaller the effect the loss, or addition, of a neutron has on the overall FOS density gradient. In fact except for the very first few of the lightest elements, the effect due to variations in an atoms neutron numbers upon its outer chemical gradient, and thus its overall potential chemical activity, is almost "negligible". This is because the addition of a neutron brings about little change in the outer gradient where the last of an element's chemical electrons reside, due to the limited effect of a neutron's own neutralizing electron's reach, as compared with that of the addition of a proton to the nucleus. But even with the addition of another proton, and additional nuclear stabilizing "neutrons" the characteristics from one atom to the next are usually not all that different. Except again for the lightest elements for which the addition of a couple or more nucleonic bodies to their nuclei results in a fairly significant change in their overall FOS density gradient, and thus their electrochemical characteristics. Hydrogen does not just gain weight by the addition of one or two neutrons. Its chemical characteristics[13] change enough that heavy water using deuterium is actually toxic. This kind of behavior is not predicted by the more popular models, and their models predict that the extra neutrons should have little effect. In fact, the effect should be a weaker behavior from neutron interference, which should not happen at all by some models, and in fact, the opposite happens and the strength by which electrons are held is quite significant for hydrogen. Affecting not only some of its physical properties but also its chemical properties such as its interaction with oxygen and thus the properties of water. Including making water toxic.

---

[13] Data from Encyclopedia Britannica

Atomic data of the three main hydrogen isotopes.	hydrogen	deuterium	tritium
atomic mass [u]	1.007825	2.0140	3.01605
natural abundance [%]	99.985	0.015	» 10^{-18}
half-life time in years in this case [tritium is the first unstable isotope of hydrogen]	stable	stable	12.26 yrs
ionization energy [eV]	13.5989	13.6025	13.6038
thermal neutron capture cross section [10^{-24} cm^2]	0.322	$0.51 \cdot 10^{-3}$	$<6 \cdot 10^{-6}$
nuclear spin [h/2p]	+½	+1	+½
nuclear magnetic moment, nuclear magnetons [μN]	+2.79285	+0.85744	2.97896

Table 6- atomic data of the three main hydrogen isotopes

Some physical properties of molecular hydrogen, hydrogen-deuterium, and just deuterium.	hydrogen	hydrogen deuteride	deuterium
gram molecular volume of the solid at the triple point (cubic centimetres)	23.25	21.84	20.48
triple point (K - Kelvin)	13.96	16.60	18.73
vapour pressure at triple point (mm Hg) (millimetres of Mercury)	54.0	92.8	128.6
boiling point (K - Kelvin)	20.39	22.13	23.67
heat of fusion at triple point (calories/mole)	28.0	38.1	47.0
heat of vaporization (calories/mole)	216 (At 20.39 K)	257 (At 22.54 K)	293 (At 23.67 K)

Table 7- molecular properties of the first two hydrogen isotopes

Physical properties of the different types of water made with different isotopes	hydrogen oxide	deuterium oxide	tritium oxide
density at 25 degrees Celsius in grams per milli-liter (ml)	0.99707	1.10451	—
melting point, in degrees Celsius	0	3.81	4.49
boiling point, in degrees Celsius	100	101.41	—
temperature of maximum density, in degrees Celsius	3.98	11.21	13.4
maximum density in grams per milli-litre (ml)	1.00000	1.10589	1.21502

Table 8- physical properties of water made from the three main isotopes of hydrogen

With the progression to the third element, lithium ^7_3Li, of the periodic table, we again find ourselves dealing with a valence shell occupied by a single electron, which can be readily shared with another atom. The main difference between this second shell and the first one is the surface area or volume of FOS here for an electron to cover. In comparison, it is quite a significant increase in area for an electron to cover, which results in this area being capable of holding up to eight electrons and no more. For the simple reason, that eight electrons can cover the entire "area-volume" of this shell-region and keep it neutralized via their wakes - which limit the number of electrons by limiting the areas of the gradient which are at rest long enough to return to a reasonable conductive state. The wakes of each lower consecutive sub-shell have not yet had the time to expand to prevent a region of FOS capable of supporting another electron from taking up residence here in the "L" shell. However, the lithium atom's gradient is too weak to hold on to an additional seven electrons in its second shell (L). Its third defining electron is held onto with a strength of 5.392 eV, and it is capable of holding onto an extra fourth electron with a strength of 0.618 eV. The limiting factor preventing it from holding onto additional electrons is simply being the strength of the FOS density gradient.

With hydrogen and lithium, we are looking at the first two chemically active elements of the periodic table whose valence shells permit them to accept, share or give up their outer electron to varying degrees, because of the "weakness" of their gradients at the level of their valence shells. In elements whose valence values are usually negative [Check.] the outer gradient structure is great enough to strongly attract and hold additional electrons. For instance, oxygen by definition should be neutral when eight electrons are in orbit around its nucleus

- with two in the 'K' shell and six electrons in the 'L' shell. However, because of the strength of the FOS gradient and the resulting surface area and volume in forming the shell - six electrons are inadequate to neutralize this reasonably conductive body of FOS forming this shell. Thus, two more electrons are likely to become "trapped" around any oxygen atom. And only two for the reason that beyond the potential orbitals of two more electrons the dissipating wakes of all eight electrons prevent any region of FOS forming that can continually hold onto more than two additional electrons, but perhaps, more importantly, the combined "disturbance" of the expanding wakes of all the electrons now produce a wake dissipation region that electrons are not attracted to and in fact is a barrier to them. This barrier exists up until the energy of the wakes have themselves dissipated enough to once again allow the FOS to form a better conductive region. The formation of standard shells that hold the same number of electrons regardless of the nucleus is based upon this area/volume-density relationship.

Electronegativity is a great way to look at the condition of the outer shell's density gradient. Electronegativity is a measure of an atoms ability to attract electrons. And the Pauling scale is the most commonly noted on tables for indicating an atoms state of attracting additional electrons or giving them up. Fluorine is the most electronegative of the elements and is assigned a value of 3.98. [The idealized value is often cited as 4.0] While francium is the least electronegative and has a value of 0.7. Atoms like cesium and francium, and their kin on the left side of the periodic table, are highly reactive because they yield their electrons easily, while atoms like fluorine and chlorine, and their kin on the right side, are highly reactive because of their strong tendency to capture electrons.

[More on this in the next edition of the book.]

The Death of the Dark Energy Idea

Periodic table of electronegativity using the Pauling scale
Atomic radius decreases => Ionization energy increases => Electronegativity increases

	1	2	3	4	5	6	7	8	9	10	11	12	13	14	15	16	17	18
1	H 2.20																	He
2	Li 0.98	Be 1.57											B 2.04	C 2.55	N 3.04	O 3.44	F 3.98	Ne
3	Na 0.93	Mg 1.31											Al 1.61	Si 1.90	P 2.19	S 2.58	Cl 3.18	Ar
4	K 0.82	Ca 1.00	Sc 1.36	Ti 1.54	V 1.63	Cr 1.66	Mn 1.55	Fe 1.83	Co 1.88	Ni 1.91	Cu 1.90	Zn 1.65	Ga 1.81	Ge 2.01	As 2.18	Se 2.55	Br 2.96	Kr 3.00
5	Rb 0.82	Sr 0.95	Y 1.22	Zr 1.33	Nb 1.6	Mo 2.16	Tc 1.9	Ru 2.2	Rh 2.28	Pd 2.20	Ag 1.93	Cd 1.69	In 1.78	Sn 1.96	Sb 2.05	Te 2.1	I 2.66	Xe 2.60
6	Cs 0.79	Ba 0.89	*	Hf 1.3	Ta 1.5	W 2.36	Re 1.9	Os 2.2	Ir 2.20	Pt 2.28	Au 2.54	Hg 2.00	Tl 1.62	Pb 2.33	Bi 2.02	Po 2.0	At 2.2	Rn 2.2
7	Fr 0.7	Ra 0.9	**	Rf	Db	Sg	Bh	Hs	Mt	Ds	Rg	Cn	Uut	Fl	Uup	Lv	Uus	Uuo

*	La 1.1	Ce 1.12	Pr 1.13	Nd 1.14	Pm 1.13	Sm 1.17	Eu 1.2	Gd 1.2	Tb 1.1	Dy 1.22	Ho 1.23	Er 1.24	Tm 1.25	Yb 1.1	Lu 1.27
**	Ac 1.1	Th 1.3	Pa 1.5	U 1.38	Np 1.36	Pu 1.28	Am 1.13	Cm 1.28	Bk 1.3	Cf 1.3	Es 1.3	Fm 1.3	Md 1.3	No 1.3	Lr 1.3

Figure 39- Electronegativities of the elements
For more on the Electronegativities of the elements – see Wikipedia

Electron affinity can show us...

Periodic table of electron affinities

	1	2	3	4	5	6	7	8	9	10	11	12	13	14	15	16	17	18
1	H -72.8																	He >0
2	Li -59.6	Be >0											B -27.0	C -121.8	N >0	O -141.0	F -382.2	Ne >0
3	Na -52.9	Mg >0											Al -41.8	Si -134.1	P -72.0	S -200.4	Cl -348.6	Ar >0
4	K -48.4	Ca -2.4	Sc -18	Ti -8	V -51	Cr -65.2	Mn >0	Fe -15	Co -64.0	Ni -111.7	Cu -119.2	Zn >0	Ga -40	Ge -118.9	As -78	Se -195.0	Br -324.5	Kr >0
5	Rb -46.9	Sr -5.0	Y -30	Zr -41	Nb -85	Mo -72.1	Tc -60	Ru -101.0	Rh -110.3	Pd -54.2	Ag -125.9	Cd >0	In -39	Sn -107.3	Sb -101.1	Te -190.2	I -295.2	Xe >0
6	Cs -45.5	Ba -14.0	* -45	Hf >0	Ta -31	W -79	Re -20	Os -104.0	Ir -150.9	Pt -205.0	Au -222.7	Hg >0	Tl -37	Pb -35	Bi -90.9	Po -180	At -270	Rn >0
7	Fr	Ra	**	Rf	Db	Sg	Bh	Hs	Mt	Ds	Rg	Cn	Uut	Fl	Uup	Lv	Uus	Uuo

Figure 40- Electron affinity of the elements
For more on Electron affinity – see Wikipedia

The reason we first looked at electron-proton interactions is that the basis for gravity is based upon the same interactions between electrons and the FOS gradients of any nuclear mass. They vary only in the degree of their strength of attraction. When we looked at electron quota levels, we were concerned with what number of electrons a gradient could attract and hold. <u>Gravity is only quantitatively different from the way electrons interact with a positively charged body.</u> Now what we are interested in is gradients that can weakly attract but not hold onto an electron. A "gravitational field" is only the accumulative left-over gradient forming fabric of usually a considerable mass of atoms. Or in other words, the combined energy of compaction of the surrounding fabric around a group of atoms. Every atom possesses a gravitational field, which is just the remaining un-neutralized FOS density gradients produced by all atoms. Regardless of the surrounding number of electrons (see figure 26b). Thus, as with gravity, the electrons are still attracted to FOS gradients in the same manner, although weakly, but now they are attracted to not one but usually trillions upon trillions of atoms. Together they produce a weak but immense and far-ranging density gradient. Which started out as the "too weak to hold an electron at all gradient" about an atom, but able to attract other electrons when this extra fabric is added to many others. A gradient, which has no singular point of origin in space, but instead stems from a general region of space.

Whole atoms and not just their electrons are drawn deeper into the density gradient field, because even though the electrons are far more strongly attracted to being part of an atom their favoritism of one side of the atom induces the entire atom into moving in the direction of the field's center by generating a pressure differential. Thus, if the electrons' host atom is free to move, then the combined effort of the most weakly held electrons within a given atom shall draw their host nuclei towards the most prominent gradient. Accomplishing this task regardless of the fact that they are bringing together bodies (nuclei) who would under their own influence repel away from each other.

For example, in the atoms of let's say of a meteor approaching the earth, the electrons are very weakly drawn towards the FOS gradient of the mass of the earth. They do not individually make a straight path for it. Instead, what happens is that they find the side of the atom facing the earth slightly more conductive, and thus extend their orbits & the time they spend on this side of the atom. As a consequence, the wakes of the electrons are now more prominent on one side of each of the atoms of the meteor, and thus the nuclei (and thus the meteor) are forced by this pressure differential towards the earth. In the end gravitational attraction between two masses at the atomic level is simply a function of the electron distribution around their parent atoms. Once they begin to favor one side, they can move their parent nuclei in the direction of the gradient they are responding to. A very important point.

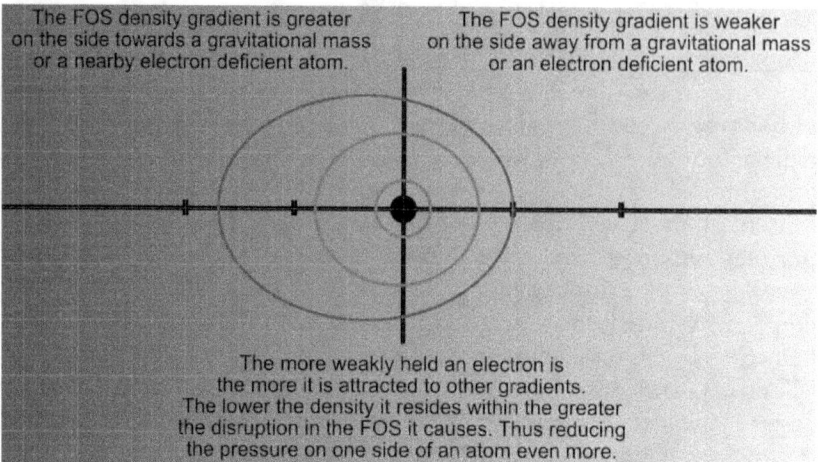

Figure 41- The more weakly held electrons are more sensitive to FOS gradients

This century's presently most accepted theories on the universe have no structural mechanism for how gravity does what it does, and when it comes to positive charge the most widely accepted model is that positive charge is the result of two up-quarks each producing a charge of positive 2/3's and a third down-quark with a fractional charge of negative 1/3. Thus, with their charges adding up to one unit of positive charge. With the quarks being held together by gluons. This description of the most popular model does not make one say - "Ah, so that's why an electron is attracted to a proton." This statement to which we are referring to has no substance. It doesn't say anything about positive charge. Let alone how the gluon tubes keep quarks together. Nor help that with quantum chromodynamics [three versions of each type of quark – red, green, blue] there are actually nine versions of the proton and nine versions of the neutron. [See tables 2 and 3 in chapter one.] Since there has to be so many kinds – it makes one ponder exactly the odds of the right kind coming together to form the proton and neutron in their model. What would the outcome of the wrong combinations have in their model? As they cannot exist.

Gravity and positive charge are at the root of it the same "force" varying only in degree. I realize of course that even though the formulae of the supposed two different forces are structured identically, this is not a proof of their "oneness". For the formulae are only algebraic expressions used to describe the relationship between two "bodies" in relation to their distance from one another, masses and the constant which may be used to calculate the force they exert upon one another. For proof, we only have to look at the fact that in lightning strikes that negative charges from the clouds move down to the Earth while positive charges move up. And perhaps more significantly we have a mystery that exists as to why positive charges are being accelerated away from the Sun triggering detectable temperatures of up to 2 million degrees K (Kelvin) in the corona while the surface of the Sun, the photosphere, is on average only about 5,778 degrees Kelvin.

Gravitational fields and positive charge fields are one and the same, but generally, a gravitational field is on the order of 10^{-40} times weaker since it generally refers to the attraction of an electron to the material of a FOS gradient, that in reality is the accumulation of the left-over gradient forming fabric of a huge number of electrically neutral atoms. In the case of an object on the Earth, they are not attracted to the field of an individual atom, but instead to the combined fields generated by the presence of trillions upon trillions of atoms. So, the atoms that they encounter on their way to being conducted towards the "center" of this gravitational field become mere obstacles in their way. If immovable and no other path is available then those atoms that are in their way are then forced to bare the weight of those who they prevent from moving further into the FOS gradient (gravitational field) of the "attracting" mass.

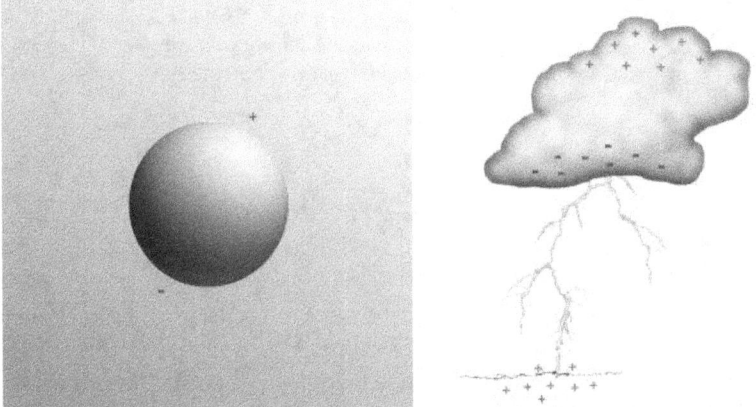

Figure 42- Atomic and cloud charge distribution within the Earth's gravitational influence

Since a proton is the initiator of this field and the electron is the effector of gravitational attraction, then it follows that a lone proton should not be attracted to a mass like a hydrogen atom would be. In lightning storms involving lightning strikes that hit the Earth, positive charges accumulate and travel up. Similarly, the Sun is a source of free positive charged nuclei who start out with a high kinetic energy near the surface of the Sun at around 5,778 degrees Kelvin in the photosphere and are seen to accelerate away into the corona where they obtain a kinetic thermal energy on the order of 2 million degrees Kelvin. And thus, are being ejected out violently like one might assume the FOS theory must indicate. It is the FOS gradient pressure that moves protons so the pressure difference from one femtometre to the next is very small but over large distances such that a proton with near-zero kinetic energy should undergo a large acceleration as is observed and is considered a mystery. No magnetic reconnection phenomenon. Magnetic reconnection is something that has never been seen in the laboratory and is adamantly refuted by many electrical and plasma engineers. Such as Donald E. Scott the author of The Electric Sky.

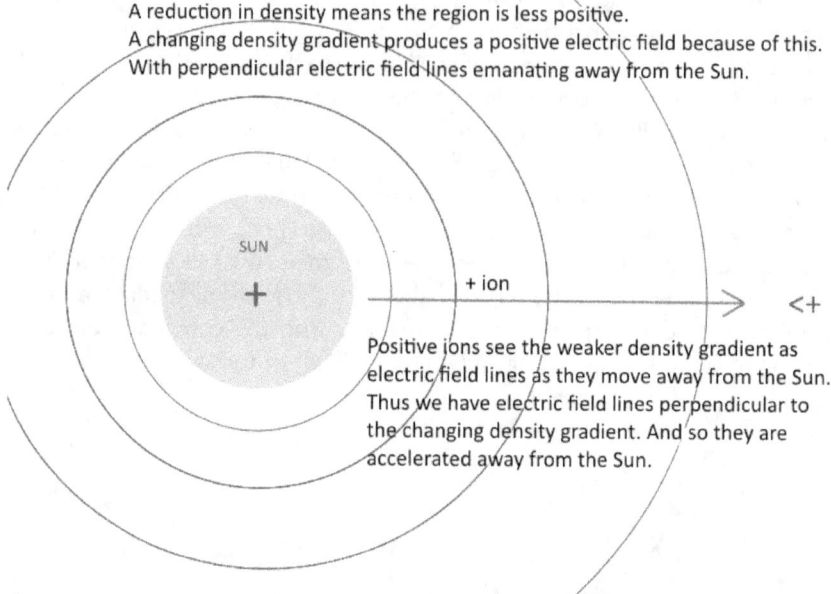

Figure 43- The Sun appears as a more positive body to positively charged ions

At the Sun, any star, the positive ions see the changing weaker density gradient as being equivalent to electric field lines as it moves away from the Sun. Thus, we have electric field lines perpendicular to the changing density gradient. And so any positive charges in such a changing density gradient are accelerated away from the Sun.

A "neutral" atom is attracted with a force based on its mass and its collection of electrons. Its electrons are what is drawing it into the FOS gradient, and thus if one begins to strip away an atom's electrons, it is less and less attracted to the Earth. Inversely of course if one adds electrons to a "neutral" atom, then it is more greatly attracted to the earth than an atom with fewer electrons. We already are well aware of the earth's potential to act as a ground, or positive attractor for free electrons. We can sum up the potential experimental proof by saying that neutral atoms or groups of atoms fall at a normal rate of acceleration while negatively charged atoms, or groups, are predicted to fall with a greater rate of acceleration, while inversely positive ions and groups are predicted to accelerate more slowly. The more positive the lower the particle's acceleration. And this is exactly what is seen. Thus, the empirical data supports the model.

If electrons are the effectors of gravitational attraction then would it not follow that as the size of the nucleus increases that the electrons closer in are more attracted to the nucleus than any external FOS gradient (and the internal nuclear electrons are shielded to some degree from FOS gradients outside). Proton mass we previously described as the result of the positrons sending out waves and triggering compression of the material around them. In effect 'pinching' themselves to the point in space they are at. It follows then that as more protons are centered around a point in forming an atomic nucleus that the waves are less effective and the protons begin to interfere with one another. And therefore, the apparent mass of the nuclei would seem to decrease - as in become less and less the sum of the masses of the constituent nucleons. As is observed. A phenomenon termed the "mass defect". The FOS Theory series explains why the mass of the nuclei do not also reflect the sum of its pions as well as any other constituent particles. In some cases, some of the so-called particles simply do not exist until a nucleus is destroyed in an attempt to try and seek out knowledge from 'particle' debris of the rubble. While also it is how mass is defined as it tries to stay in place or remain in constant motion. The energetic electrons (pions) outside the nucleus possess kinetic energy that is seen in part due to mass as momentum carries it in a straight line, but as part of a neutron/nucleus, the pion is more interested in the region around the parent proton. The mass of a negative pion is around 139.570 MeV, the mass difference between a proton and a neutron is just 1.293 MeV. In contrast, the mass of an ordinary electron is around 0.511 MeV. Part of the momentum of an electron is related to the energy, or density in the fabric maintaining the electron, but also due in part to the FOS material in the scoop or cone-like portion of the electron. This scoop of material gives it something to be attracted towards other than some electric field (gradient) placed close by. While in the nucleus that material it is attracted to is the FOS gradient around the nucleons. Only a pionic-electron is of sufficient conductive energy to be kept around nucleons within the nucleus.

A proof of the existence of the FOS density gradient about a proton is the fact that the FOS series predicts inertia for bodies possessing electrons and as well as predicting that a lone proton, and other electron-deficient nuclei, will be forced by its own gradient away from bodies like the Sun. Proof of the pairing of an electron to a proton to form a neutron, versus the quark triplet model, can be seen in the existence of ultra-cold neutrons that as they decay their wavelength increase. The FOS Theory series predicts inertia because as a nuclear mass is set into motion, the newfound compression of the fabric ahead of it or that part of the new "electric field/mass" that is formed is an increase on one side of the FOS density gradient. Thus, this new denser region draws the electrons towards it - thus also continuing to draw the nucleus which therefore maintains the existence of the newly formed "electric field/mass" for the given velocity.

[More on this in the next edition of the book.]

We can now see the relationship between gravity and quantum mechanics, simply as variations in scale and intensity of complete aether/FOS gradients or as pockets of aether/FOS density gradients. [Images here showing a gravitational gradient versus a potential electron orbital position or barrier fluctuations between gates on a microchip.]

[More on this in the future second edition of the book.]

Einstein's Principle of Equivalence and Elevators and Rockets.

Albert Einstein wanted to understand why when objects of any mass fell to the Earth, under vacuum conditions, they fell at the same rate. And during his gedankenexperiments [thought-experiments] he started considering the acceleration of stationary and moving elevators, and moved onto considering their motions within rockets in outer space outside the effective influence of any considerable gravitational mass. Eventually he realized that photons passing through an elevator on the Earth, or in an accelerating rocket ship, would also have to experience such a displacement as well.

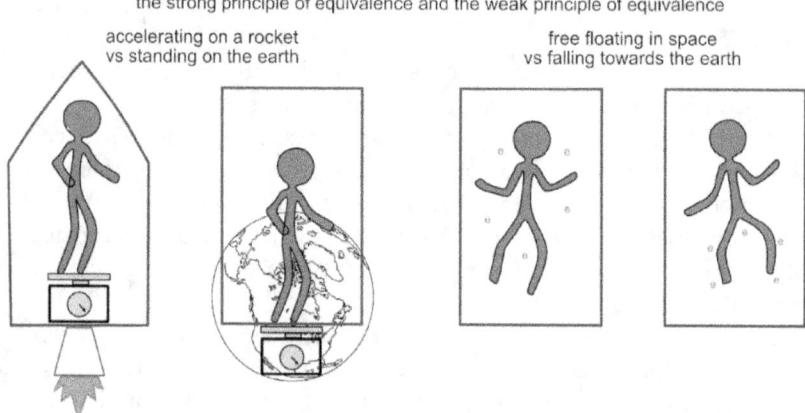

Figure 44- Strong and weak principles of equivalence

For the weak principle of equivalence, Einstein realized that there was a way to determine if the masses were freely falling towards a planet or just floating in space. That difference for at least a pair of falling bodies is that as they get closer to the planet, any pair of bodies/ masses would start to move towards one another without their own gravitational attraction to each other being the cause. This was due to their purely geometrical relationship of moving towards a common center. However, without such a secondary reference mass, there was no way to know. But now there is with this new Fabric of Space, aka FOS, model. Weakly held electrons know.

Weakly-held electrons around masses know when they are near another mass, and their preference to move towards such bodies can be detected.

The Death of the Dark Energy Idea

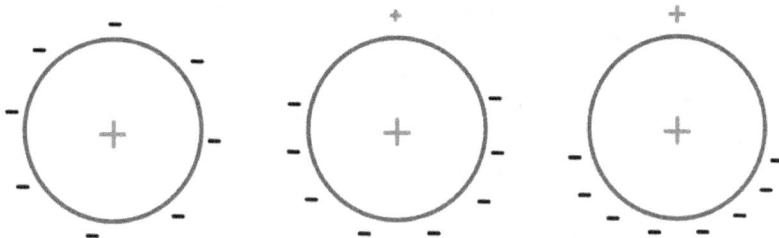

the weakly-held electron distribution pattern is affected by how close one mass is to another mass and of course, the magnitude of the 2nd mass. a similar distribution pattern can also be seen in the electron distribution around an accelerating mass

Figure 45 - weakly-held electron distribution changes

As masses with weakly held electrons approach another mass, or enter the gravitational field of a more significant mass, the distribution pattern of the weakly held electrons changes to spending more time on the side of the masses between each other. Even the electrons held by both chemical and electrostatic bonds are influenced as the aether gradient fields between bodies is a better conductor and the electron wakes create more of a pressure drop due to the electrons spending more time here.

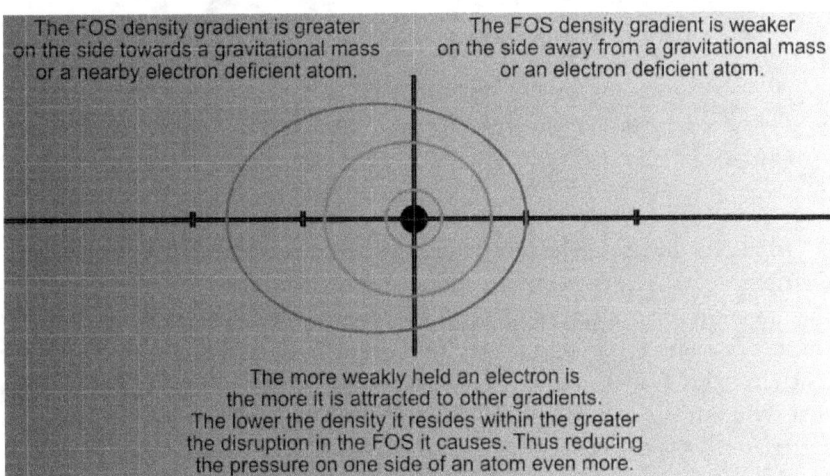

Figure 46 - the more weakly held electrons are more sensitive to FOS gradients

From the FOS model, there is also a difference for the strong principle of equivalence in that next to the Earth all the objects are moving towards the Earth due to the electron distribution around them due the electrons favoring the planet side in their orbital motions because they are within a planet's mass/aether gradient. While for the accelerating rocket the electrons would have a more normal distribution, with more of a drag on their positions due to the acceleration, and only renormalize their electron distribution as their masses adjusted to the new induced velocity. Likewise, in the weak principle of equivalence, the masses falling towards a gravitational center would have their electron orbital patterns reflect their attraction to the planet, while that in space far away from any significant influence the electrons would be for the most part evenly distributed around their atoms.

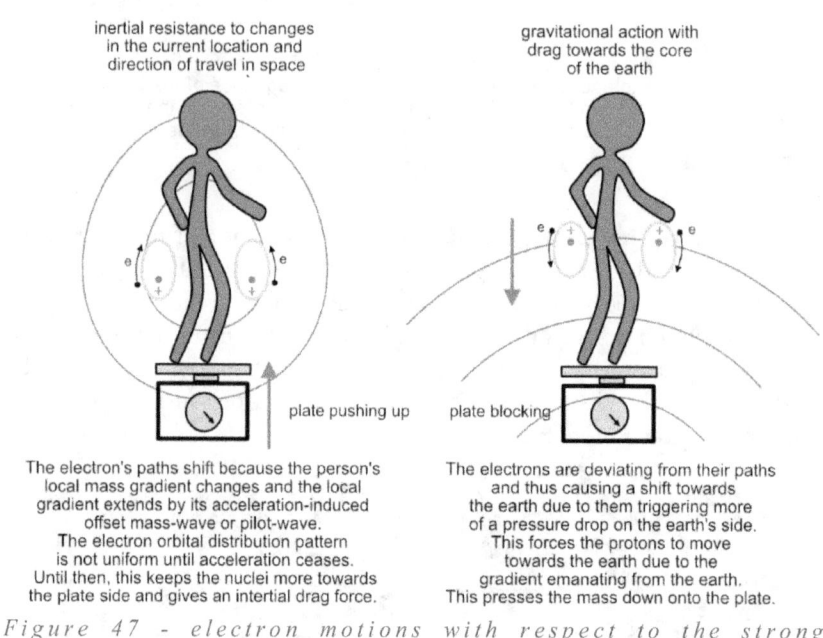

Figure 47 - electron motions with respect to the strong principle of equivalence

In bodies under acceleration, the electrons deviate from their regular orbital distribution pattern towards the induced mass wave or pilot wave, forming in the direction of acceleration. The mass's inertial resistance, which is mostly the nuclei's resistance, is in part related to the electron's orbital distribution pattern. The electrons' wakes, just as in atomic orbitals, trigger a lower pressure region that then attracts the protons, but until this changes enough, the whole mass system's resistance to change applies pressure against the source of acceleration. Similarly, the electrons around somebody resting on the Earth are deviating towards the gradient of the planet, and once again, the wake dissipation regions attract the nuclei of the body to the Earth. This is experimentally verifiable, providing evidence of an aether gradient, and in the case of the person being in the Earth's gravitational field, the electron's movement towards a person's feet is already known. So even in freefall, we can show the same thing: while a person floating in uniform space between planets, the electron distribution around their bodies is uniform. Weakly held electrons have no preferred position. And, in fact, they would prefer to stay away from any others of their kind.

The instability of the formation of Black Holes

Let us first acknowledge that it was the electrical engineers and their allies who first predicted that a pair of 'objects' should be found at the cores of some galaxies. "In the form of two interacting galaxy spanning Birkeland electrical currents with plasma clouds trapped in parallel magnetic filaments can simulate evolving galaxy formations, without the need for dark matter and black holes!" Black holes, in the model presented within these pages, cannot form because any mass exceeding the density of positrons and protons is torn apart because of the standing wave energy limits the density and size of the constituent particles of matter, and does so by reducing such attempts back into its constituent base of positrons [gamma-rays and neutrinos], or a complete reversion back into gamma-rays.

As of the release on Wednesday, April 10 2019 of the first supposed image of M87's supermassive black hole captured by astronomers from the Event Horizon Telescope team, there had been no direct evidence for black holes. Has anything changed? In fact, the image also supports the electrical model of galaxies, and what some are calling a black hole the plasma and electric universe proponents, like Wal Thornhill from the Thunderbolts Project, are saying that this image is more in line with their predicted electrical phenomenon at the cores of galaxies where giant electromagnetic forces are at work. They state this is in fact a plasmoid structure at the core of M87. A plasmoid is a semi-stable structure of an electrically active plasma and its' associated magnetic fields.

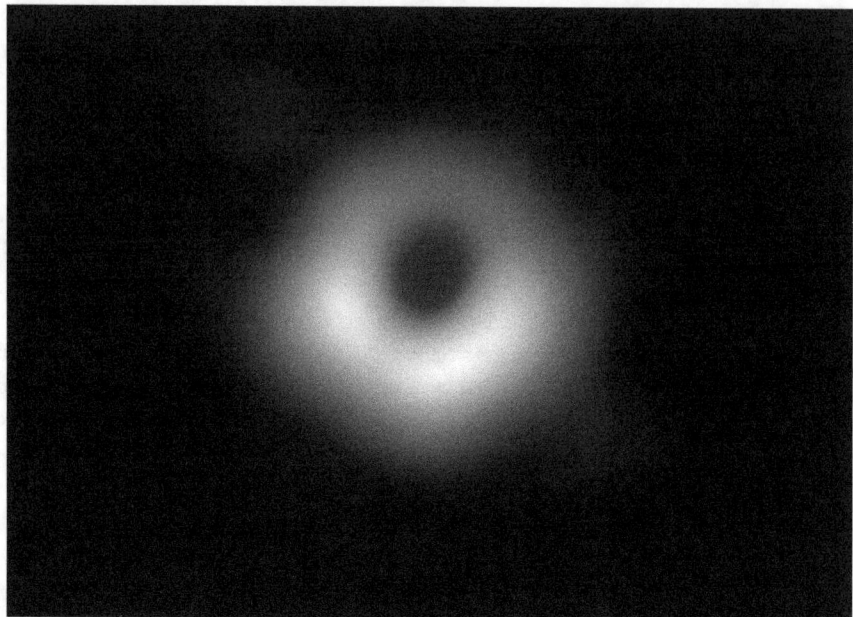

Figure 48- 2019 astronomical image of a black hole or electrically based plasmoid?
[Credit: "Black hole" imaged by the <u>Event Horizon Telescope</u> Collaboration]

The other evidence that at first seemed to confirm the existence of black holes was the motion of the stars at the core of our galaxy. Most notably by the observed motion of the Sagittarius A-star and its neighbors. It would seem that there is no optical indication of any of those stars passing behind this supposed black hole at the core of our galaxy or we would see an optical spreading of their photons. But then the time required to capture such an event may be too short based on the data collection times required. But something more important is that there are large magnetic tendrils, predicted by the electric model, that have been photographed emanating from our galactic core. Massive magnetic fields imply massive electric currents. Remember that the difference in strength between electro-magnetism versus gravity is 10^{39}! We are talking about a 1 followed by 39 zeros! Stars are giant plasma bodies. That is a star is a giant massive charged body, and should be affected by any large currents at the core of our galaxy.

It does not even make sense to talk about how the more massive the black hole, the more matter and energy are able to escape from them, or their accretion disks. This defies logic on any level, and contradicts the very definition of black holes, regardless of the physics of accretion disks, and against what we would expect from black holes based on their definition. And they just say look at the core of our galaxy or several others and the proof is staring at you right in the face. No. The magnetic fields exist because of large electric fields. No experiment shows a magnetic field without an electric field. Bar magnets, and their ilk, have magnetic fields not because of their mass but instead because electrons can take up fairly stable roughly single planar orbits – inducing magnetic fields. Like an electromagnet.

Links
http://news.discovery.com/space/galaxies/monster-magnetic-lobes-milky-way-core-130128.htm

Gravity and the formation of solar systems

One of the key problems with a gravitationally solely based formation of solar systems, without consideration of the key role that electricity plays, is that there is no mechanism of shedding excess momentum as the solar system forms. You only have to stump or silence those who only consider gravity, and/or dark matter, is by asking them how a spinning accretion disk loses its angular momentum? [The Electric Sky by Donald Scott]

Perseus Cluster – is supposed to have a x black hole that has the effect of creating bubbles of superheated gas. The black hole is blasting away matter from the center of the galaxy. Where each bubble is supposed to be the size of the Milky Way. This is supposed to show the corresponding relationship between the size of a galactic black hole and the mass of its surrounding galaxy. The ratio of the two according to x is on the order of a grape and the mass of the Earth. As the black hole begins to devour matter so it starts to pour out energy. Like a cosmic broom. That energy sweeps matter back out from the center of the galaxy preventing it from clumping together to form new stars. [Television show: Horizon: Swallowed by a Black Hole [2013-06-26]]
[More on this in the future second edition of the book.]

Quasars are now being considered astronomical objects that are feeding black holes. One more step of quasars moving towards being accepted as electrical objects.
[More on this in the future second edition of the book.]

Chapter Four

THE NEUTRON AND THE NUCLEUS

Once the problem is eliminated by an excuse, there is no need to reflect upon it any more.
– Erwin Schrödinger (1887-1961)

I know that most men, including those at ease with problems of the greatest complexity, can seldom accept even the simplest and most obvious truth, if it be such as would oblige them to admit the falsity of conclusions which they have delighted in explaining to colleagues, which they have proudly taught to others, and which they have woven, thread by thread, into the fabric of their lives.
- Leo Tolstoy (1828-1910)

The Neutron and the Nucleus

In the FOS model series, neutron characteristics and structure are derived in essence from the presence of and an increased degree in the intensity of the disruption caused by a pionic-electron's wake near a proton as it is conducted around it. The increased disruption is a combination of two factors; one related to the density of the FOS formed when so close to the proton, and also the drop in gradient pressure generated by the pionic-electron's wake now being so close to the proton. The wake causes an aether density gradient disruption, triggering the lowering of its density which translates into a momentary decrease in the pressure on that side of the proton. Precession of the electron around the proton is triggered in the same manner as the electron quantized orbitals of the atoms, because the electrons are conducted into the areas where gradient reformation is more underway. The denser a region, the better a conductor it is and the better it is at bringing the electron into that region. With the coverage achieved by the induced precession neutrality is gained by the electron reducing the density gradient around the proton. The reduced density gradient deters the conduction of other electrons, thus giving rise to the neutronic systems neutral charge. No other electrons are attracted to the core proton, and it presents no significant gradient to other protons to repel them. Thus, the pionic-electron's relationship to this neutron's core proton is similar in nature to their structural behavior in the formation of the electronic shells around an atom. [Remember though outside a nucleus a neutron is not stable and breaks up in about fifteen minutes.] If a neutron gets close enough to another proton the negative pion's wake can permanently unite the two of them to form the heavier version of hydrogen called deuterium. xx

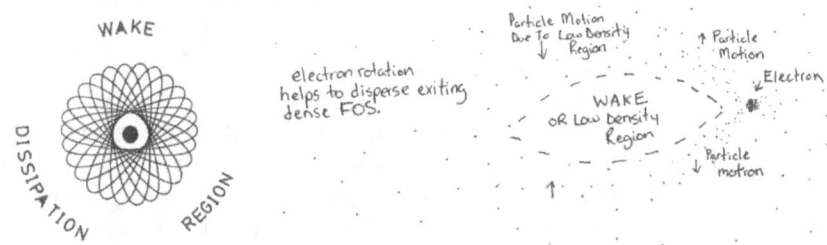

Figure 49- Wake dissipation action by electrons creates quantized regions

[Insert image of wake region in a deuterium nucleus. Drawing in progress.]

Previously we discussed the mass of a proton, and that of a positron, with the mass of the neutron being somewhat greater than that of the mass of a proton plus that of an electron. Remember we proposed that mass is in effect the anchoring resistance that matter has either its desire to move in one direction or to remain stationary. It would therefore seem then that the neutron mass is the result of the enhancement of the proton's mass with the electrons' motion also effectively aiding to keep it in place. Which explains why the neutron's mass is greater than just the addition of the proton mass plus the electron's mass.

Pions [negative, positive and neutral] were first proposed in a paper by Hideki Yukawa in 1935 as the mediator of the strong force. He was awarded the Nobel Prize in physics for his prediction of the pi-meson and subsequent discovery by Cecil Frank Powell, Giuseppe Occhialini and César Lattes in 1947.

Outside the nucleus, all pions quickly break down into their fundamental forms of either electrons, positrons or gamma-rays. And when we say quickly, we are talking about them having a mean lifetime on the order of around 26 nano-seconds (2.6×10^{-8}). First decaying into muons then the muons changing with a mean lifetime of about 2.2 microseconds (2.2×10^{-6}) finally into electrons or positrons. Using the high energy electrons, which are normally referred to as negative pions, as the effectors of the strong force also instantly resolves something called the proton-neutron distribution* problem of nuclei. That is that there is pretty much no way to distribute the neutrons to prevent the formation of very large repulsive forces between adjacent protons in nuclei. But if we instead model the neutron as a pairing of a proton with a nuclear electron, then there is no distribution problem and the pionic electrons are simply ushered into the spaces between protons that are beginning to build up too much of a positive charge/ gradient. The idea of electrons turning into negative pions as the initiators of the strong force is not a new idea, but it is one that was dismissed as a possibility early in the 20th century. The key evidence at the time being that high energy electrons in the form of muons were never seen entering nuclei, and thus the concept was tossed aside into the junk heap of ideas. These people did not have all the data we have now, and others in the mainstream have simply not bothered to revisit this concept in light of the new empirical data. [Not just one piece of data but hundreds of them.] The original idea of electrons as the effectors in forming neutrons proved to be a very short-lived part of a theory, which is still an important theory of beta radioactivity, by Enrico Fermi. This idea of electrons being the quanta responsible for keeping protons and neutrons together to form nuclei was dismissed partially on the experimental evidence that if a gas of nucleons was bombarded by a stream of electrons - "little interaction takes place". "Because of the wave structure of the electron, it can be shown that an electron inside a nucleus would have to move about with so much kinetic energy that it could not be held there by known forces." Another flaw in their logic. Assuming they understood charge at all within the confines of the nucleus. "Since the nucleus is very tiny, the position of the electron inside the nucleus would be so accurately known that its momentum would, according to the uncertainty principle, be extremely large and hence its kinetic energy would be very large." Also supporting this conclusion was the fact that the electron wavelength was greater than the diameter of the nucleus. So many have stated;- "How could an electron even fit within the confines of a nucleus." I do not doubt the results of the experiment, but I do disagree with the overall conclusions of the experimenters and those who agree with them. They are completely wrong based on one simple fact. Do electrons enter nuclei? Absolutely! Electron Capture, or EC [formally Electron Orbital Capture – EOC], which is one of the most common forms of radioactivity. Normally ordinary electrons cannot react with nucleons or enter a nucleus of an atom unless that atom has too few neutrons in relationship to its number of protons. And also, we have data on electrons coming out of nuclei in the form of negative beta-rays. Carbon 14 into Nitrogen 14 as an example. So, they are within nuclei, and no one can dispute this without denying modern nuclear physics, but they are not in their normal form (conduction strength & structure). The particles in question are the tau, pion, muon and electron. The former two decay into electrons. On top of that if you use the Classical Electron Radius equation, you simply have to substitute the mass of a pion to have the electron become considerably smaller than the radius of a proton. Again, the data supports this.

For some time now we have known that in a nucleus that is too proton-rich (or neutron deficient) an "ordinary" electron can be drawn into the nucleus of an atom from the 'K' shell - the first orbital. This form of radioactivity, as it is known from the release of energy from this process, is called Electron Capture [aka Electron Orbital Capture] or just EC. And this type of radioactivity occurs in quite a few isotopes of the elements. It happens within a range of time from fractions of a second to minutes, hours, days, years, decades, centuries,... If the reaction is weak, then an electron simply enters the nucleus, and the proton count goes down, and the neutron count goes up. If the reaction is above a certain energetic threshold, then an electron enters the nucleus but a positive beta particle, positron, is ejected. In this case, it seems the electron triggers in its jump a gamma-ray event that reacts with the gamma-ray dense material around a nucleus to trigger the formation of a positron – but without triggering the formation of another electron. This is proof that there is an exception to the conservation of charge. Scary? Alternatively, a nucleus that is too rich in neutrons can eject a negative beta particle – an electron.

A small sample of just a few nuclei that experience EC. From the Table of Isotopes from the CRC Handbook of Chemistry.

Radioisotope	Half-life	Radioisotope	Half-life
^{7}Be	53.28 days	^{56}Ni	6.10 days
^{37}Ar	35.0 days	^{67}Ga	3.260 days
^{41}Ca	1.03E5 years	^{70}Ga	21.1 mins
^{44}Ti	52 years	^{68}Ge	270.8 days
^{49}V	337 days	^{72}Se	8.5 days
^{50}V	1.4E17 years	^{77}Br	2.376 days
^{48}Cr	21.6 hours	^{76}Kr	14.8 hours
^{51}Cr	27.7 days	^{94}Ru	52 mins
^{53}Mn	3.7E6 years	^{94}Pd	9 secs
^{54}Mn	312.1 days	^{111}Xe	0.7 secs
^{57}Co	271.8 days	^{129}Cs	1.336 days

Table 9- Isotopes that can experience Electron Capture

The Classical Electron Radius is often cited as proof that an electron cannot enter a nucleus simply because it is too big. But here again, they would be wrong because the equation relies upon the mass of an electron in the denominator of the equation. The fatal flaw being that muons are considered just energetic electrons with a mass of ~207 times greater than that of an electron. And pions decay into muons. The rest mass of a negative pion is ~273 times that of an electron.

$$r_e = \frac{1}{4\pi\varepsilon_0}\frac{e^2}{m_e c^2} = 2.8179403227(19) \times 10^{-15} m$$

Classical Electron Radius: $r_e = 2.8179403227(19) \times 10^{-15}$m
Proton charge Radius: $r_p = 0.8751(61) \times 10^{-15}$m
Radial difference ~ 3.2199 or 3.22 times larger than a proton
Using instead
Muon rest mass yields: $r_{mu} = 1.3589 \times 10^{-17}$m
Radial difference ~ 64.402 or 64 times smaller than a proton
Pion rest mass yields: $r_{pi} = 1.0305 \times 10^{-17}$m
Radial difference ~ 84.926 or 85 times smaller than a proton

[Proton radius compared to electrons, muons, pions]

Proton	Electron	Muon	Pion
0.8751×10^{-15}m	2.8179×10^{-15}m	1.3589×10^{-17}m	1.0305×10^{-17}m
	~3.22 x larger	~64 x smaller	~85 x smaller

Table 10 - Proton radius compared to electrons, muons, pions

Now someone might want to point out that muons have never been observed entering nuclei in experiments designed to observe this and thus cannot be the source of the strong force. Actually, if you break matter in experiments with antimatter, it is negative, positive, and neutral pions that you end up producing. Even so back to the muon argument – it comes back to the failure to realize that only neutron-deficient nuclei which can accept/force electrons into them. That is the essence of the form of radioactivity called Electron Capture. As all other balanced and thus stable nuclei possess a nuclear-potential barrier that is in effect like the full electronic shell of an atom. If it is occupied, not of the correct density gradient nature, then such a region around a nucleus will not accept another electron. Regardless of its energetic state of being either a muon or a pion. Muons are not the source of the strong force it is the more energetic form of electrons called the pions that are.

In Electron Capture the momentum and the kinetic energy is a factor here, but only because the structure of the electron does not permit it to possess a greater attraction towards the nucleus in its current state. An increase in the strength of the FOS gradient must be of a value that is great enough to force a structural related conduction change in an electron to overcome its momentum to move in any closer. A change that if possible, for some isotopes for most elements, will take some time and this time will vary depending on the isotope. Beryllium-7 is the first such nucleus with a half-life of about 53.28 days. The next possible isotope that might decay by EC is Carbon-11 with a half-life of only 20.3 minutes. There is a lot of empirical data for this form of radioactivity that is called Electron Capture – EC. [Formerly called Electron Orbital Capture or EOC.] How common is this form of radioactivity? If you get a copy of the CRC Handbook of Chemistry and Physics and look at just the first 54 elements of the periodic table, up to and including Xenon, there are 251 isotopes in which Electron Capture can be a form of radioactivity. "Can be" because some isotopes can also decay by other means. The other most common forms of radioactivity being beta electron emission and beta positron emission.

So, ordinary electrons are known to enter the nucleus under the correct conditions. But only under the correct conditions. If given enough time around the proper nucleus, too proton-rich, one of the electrons will inevitably enter into such a nucleus. Around the nucleus, there also exists a wake dissipation region from the nuclear electrons. This would appear to be the same thing as that which triggers the formation of electronic shells. But just on a different order of magnitude.

Even electrons that have been excited by the use of a cyclotron and directed towards nuclei will normally encounter a layer of fabric, which will only be able to conduct these electrons by ordinary means (energies). The momentum of such electrons would find themselves too great to be properly conducted by the generated gradient to overcome any nucleus' nuclear-electron's wake barrier, a relatively poorly conducting gradient, and therefore unable to enter such nuclei. The only way for an electron to take up residence within the nucleus would be for the nucleus to be neutron deficient. So that its wake dissipation region would be weaker, the gradient stronger, and then for the electron to have more time about the nucleus to change to a form that would allow it to enter and become part of the nucleus. Consider the time given to the electron already in the 'K' shell in ordinary Electron Capture for Beryllium-7, $^{7}_{4}Be$, with a half-life of 53.28 days at the end of which it becomes regular old Lithium-7, $^{7}_{3}Li$. The most common mode of time being between days and years for this to occur. With a few happening in seconds to fractions of seconds.

The negative pion is the excited state of an electron that will readily react with a nucleon and particularly with nuclei. Because it has already undergone the structural change of greater conductibility, and FOS displacement, by having just come from the interior of a nuclear FOS gradient. The pion (negative), short-form for pi-meson, is not considered an excited state of an electron, and yet it decays into one, but another particle that was once considered a meson, is now only considered an excited electron - the muon. Now only called a muon [aka mu-meson] since the term meson is now confined only to the quanta for the nuclear force. [What is the distinction between the strong and color force? More on this in the future second edition of the book.] Muons are said not to be able to enter a nucleus. But for more or less the same reason, the other experiments failed. You need to have them interact with proton deficient nuclei, and they have to have enough time. A muon will briefly take up residence about a nucleon or nucleus, but will not normally interact or react with them. A muon decays into an electron by double neutrino emission to become an electron. Thus, it has long been accepted that "a muon is merely an excited state of an electron." A pion decays into a muon (in about 2.6×10^{-8} seconds) by the emission of a neutrino. Now I don't understand why this has not been more vocally considered but since a muon decays into an electron by the emission of two neutrinos and is considered an excited state of an electron, and a pion decays into a muon by the release of a single neutrino - does it not follow then that a pion should be considered just an even more excited state of an electron.

We have not discussed the negative Tau (-τ) particle that is one of the particles listed in the accepted Standard Model of particle physics. Like the electron, and the negative muon, it is considered one of the leptons. The negative tau has an electrical charge of that of the electron, but its mass is approximately 1,776.86 MeV/c² or ~3,477 times greater than the mass of an electron. For the electron its mass is defined in terms of millions of electron-volts at 0.511 MeV/c². While with respect to that of a negative pion (-π) which has a mass that is about 139.57 MeV/c² (~273 x the e⁻ mass), versus that of the negative muon (-μ) whose mass is around 105.66 MeV/c²(~206 x the e⁻ mass). The negative pion generally decays into a negative muon in about 2.6×10^{-8} seconds outside of the nucleus, while the negative tau generally decays into a negative pion and tau neutrino (υτ) in about 2.9×10^{-13} seconds. It can hardly be said to have existed at all. All of what we have been told about the tau particle becomes kind of a moot point when you realize the reality of its being to be nothing more than an even more energetic state of a pion from the most energetic collisions that our engineers and scientists have been able to create, and it is pions that we look to for nuclear cohesion.

Electron(e)	Muon(μ)	Pion(π)	Tau (τ)
0.511 MeV	105.7 MeV	139.57 MeV	1,776.8 MeV
1.0 e_{mass}	~ 206 x e_m	~ 273 x e_m	~ 3,477 x e_m
stable	2.2×10^{-6} s	2.6×10^{-8} s	2.6×10^{-13} s

Table 11-electron, muon, pion and tau comparison table

There are said to be three kinds of pions; a negative one, a positive one and a neutral pion. The neutral pion is said to decay into two quanta of gamma-rays; the positive pion decays first into a muon and a neutrino, and then into two neutrinos and a positron; finally, the negative pion first decays into a muon and a neutrino, and then into two neutrinos and an electron.

Their model leaves a lot to be desired. One of the models believes that a neutral pion is "tossed" back & forth between two protons and/or two neutrons to produce the strong nuclear force between these pairs of particles. While either a positive or a negative pion can be passed back & forth between a neutron and a proton to produce the strong nuclear force. This is not a model that makes one think that this must be because it makes so much sense. Not at all. Tossing pions between nuclear bodies does not make sense as logical explanation for how the strong force works since it should trigger them to move apart - not together. I do not dispute the existence of the three kinds of pions, but I do dispute this structural claim for the positive and neutral pions in their interactions with nucleons within nuclei. Especially since pions are the decay products of matter-antimatter reactions. The annihilation of a proton, with an anti-proton, yields on average two positive pions and two negative pions. The annihilation of a neutron with an anti-proton yields on average two positive pions and three negative pions. [Hans Mes et al.]

The basis of my argument is this; if three types of pions always existed within a nucleus, then the pions at times would have to compete with one another in an uneconomical & chaotic way - defying the principle of simplicity (Occam's razor). At times one of the nucleons to which a neutral pion has been privy might now find itself closer to a nucleon that was under the control of perhaps a positive pion and now might be attracted to it due to its proximity. It is the proximity of the adjacent nucleons to one another that would permit far too often the possible exchange of pions. This would not be so bad if only one type of pion was believed to exert the strong nuclear force, but with three to govern the various pair combinations the likelihood of a situation in which pions were in conflict with one another is extremely high. Too high to make a nucleus run smoothly, and thus more complex and erratic than we normally discover nature to be. If this does not create conflict, then consider if a neutral pion interacted with a different pair of nucleons than usual, which would then force the former original pairs' pion to compete with it for the same task or move onto another similar pair. Such behavior does not seem to inspire the idea of a smooth flowing system at work. If anything, the initiators of the nuclear force should behave in a manner, correlative, of the fine movements of a near-perfect Swiss watch. Or perhaps more appropriately of a skilled juggler, either way, whatever the symbolism, their behavior one would think would be based upon a forgiving and simple structure. One that can easily internally adjust itself and rebound from an erratic movement within itself and return to a normal state.

Again, I don't doubt that the positive and neutral pions can be brought into existence, but I do dispute their necessity within the nucleus as co-initiators of the strong nuclear force in conjunction with the negative pion. An experiment to verify this would be to try to unite nucleons only with positive and neutral pions. As you may well guess in the FOS Theory series, there is a viable and stable arrangement within the nucleus that involves only a negative pion. What I am proposing is not exactly the same that is now accepted or even one that has been hypothesized - except vaguely by Enrico Fermi. As we are now talking about a wake-inducing pionic-electron that disrupts nucleon gradients.

A neutron is formed when an electron enters into orbit around a proton. Electrons of normal energies, via conduction, encountering an electron-bare nucleus would normally be unable to come this close to one of its' protons before its momentum would carry it out to a higher orbit. Long before it got anywhere near a proton. With the notable exception in the case of nuclei that are neutron deficient (too proton rich) for their sizes, and thus their nuclear electron wake dissipation regions are not the formidable barriers that they normally are and thus will trigger an electron from the K-shell to transform and enter the nucleus.

However, for an electron that has made it in this close to a lone protons' FOS gradient, it is extremely conductive. Permitting it to conduct an electron well enough to keep it into such an incredibly low and energetic an orbit. Well at least for approximately 14.7 minutes. Why this many minutes? Well, because this is the half-life of a lone neutron. [Depending on the reference source you use for looking up its half-life.] The gradient of a single proton does not provide a body of FOS conductive enough to keep an electron into such an energetic orbit for very long. Two protons though is another story. Deuterium is the first isotope of hydrogen that has a neutron as part of its nucleus. And is stable.

Further evidence for the structure of a neutron can be found in ultra-cold neutrons & their polarization of which these aspects of neutrons are further support for their structure as hypothesized in the FOS Theory series. Ultra-cold neutrons are simply non-nuclei bound neutrons whose velocities are quite low in comparison to what they could be. These ultra-cold neutrons' velocities range from 2200 metres per second with a corresponding de Broglie wavelength of about 0.2 nanometres, as opposed to a nuclear radius of approximately 0.0000009 nanometres, and as low a velocity as seven metres per second with a de Broglie wavelength greater than 50 nanometres. At these wavelengths, which are also much greater than the distances between atoms, these neutrons are readily reflected off of the surfaces of bodies of matter (as opposed to their normal dimensions which permit them to readily pass through matter). Very similar in effect to the behavior of almost any spherical object that is bounced off of a smooth surface. I've hypothesized that the reason they've reached these dimensions is that as the neutrons "decay" their electrons are moving further and further away from the protons, due to their possessing greater momentum than their conduction strength, and do so until theses neutronic-protons lose their electrons entirely. At which time they can be said to have completely decayed. But in the meantime, the proton & electron of a given neutronic system still behave as one, and like a hydrogen atom can be bounced off of a solid surface. Unlike a hydrogen atom, they can be "polarized" by means of their magnetic fields. That is they can be aligned so that their poles run parallel to one direction in space. Neutrons behave as single electron-induced magnets, when alone, because of the fact that the electron is able to circle the proton in a manner that sets up a flow of FOS similar to that found within an atomic magnet (see Magnetism), but because of their orbital period the fabric has a chance to recover & thus permit a single electron to form a magnet. Its momentum may also play a role since it may not permit the electron to precess in orbit as it normally would in the search for the most conductive FOS.

Polarization is initiated by passage through or the reflection off of a sheet of magnetized material. It would also seem that more support for these conclusions about neutrons is supported by the fact that these ultra-cold neutrons have storage times, based on the fact that their radioactive decay time is considerably shorter than the observed half-life of regular (faster) neutrons. Storage taking place within specialized vacuum "bottles". Today's presently accepted theories are unable to account for this apparent shortening of the lives of these ultra-cold neutrons, but I've become confident in the reasoning that these ultra-cold neutrons are simply much further along in their decay (process) than normal more recently ejected, or more energetic, neutrons. Their state of decay being indicated by their dimensions (de Broglie wavelength) and probably their velocities. This greater wavelength cannot be accounted for by the quark model of the neutron. And thus, is the quark model's "Achilles' heel."

I have been wondering if these neutrons are slowing down because as the electrons move further, and further, away from the protons, they permit more of the FOS gradient of the protons to be exposed. Exposed as in no longer being neutralized by its nuclear-electron partner, and thus allowing the FOS gradient to react with the surrounding fabric of space. Forcing them to slow down since it takes an electron to maintain this systems inertia [?] or in other words, it takes one to prevent the FOS gradient of a proton from reacting with the surrounding FOS. Or is it just their interaction with other particles around them and passing on their kinetic energy. Of course, one might also wonder why doesn't an ultra-cold neutron simply turn into a hydrogen atom since its' electron is at a much greater distance than the first potential orbital position of a hydrogen atom. The reason is that this is no ordinary electron and as such possess too much "energy" as a nuclear- electron. [Is this pionic electron displacing more of the gradient than a regular electron? A normal electron found in this orbital position would only possess the FOS displacement capabilities of a (non-nuclear) electron of ordinary atomic gradient characteristics.] This former nuclear proton-electron pair (neutron) remain united for a time beyond normal dimensions, whereas normal electrons would fail to have influence, because these nuclear-electrons were once pions that have much greater FOS conduction and displacement characteristics. It seems to make sense that if a particle possessed greater charge then it would be able to neutralize the positive charge of a proton from a greater distance away. As well as do it more effectively than an ordinary electron closer in. Where a regular electron would also permit a second electron to take up an orbital position above it, while a pionic electron maintains the ultra-cold neutron's complete neutrality. Preventing the formation of a gradient that would attract another electron into a stable orbit. One possible explanation as to why the multiple neutrino emission of the normal pion decay, into an electron, appears to be bypassed is that the "electron" has been provided with a greater length of time to adapt to "normal" space, and thus only possesses a smaller amount of FOS to be shed near the end - in the form of a single neutrino. As well it may already have released energy in its escape from the nucleus from which it was liberated. Could it also emit the compressed fabric at an ultra-low expansion rate (extreme long wavelength type decompression)?14 Whereas if a nuclear electron (pion)

---

[14] Explore the concept of the expansion of the fabric out of the cone-like depression in a neutron decay event. That is as

is forced directly out of the nucleus it has "very little" time to adapt to normal space and thus releases its FOS in larger or more concentrated amounts. And does this since the material has little time to expand back to a normal density fabric before it escapes from the confines of the electron's cone-like depression. With the energy (wave) being carried along by compression and rarefaction ahead of the slowing & expanding electron. A pionic electron tries to adapt in less than a millionth of a second, while in neutron decay the electron has over ten minutes to escape and adapt.

The system of a proton and an electron known as a neutron is in terms of charge neutral -for the most part. That is in this low and energetic an orbit an electron, structurally adapted to the most extremely rigid & dense portion of a proton's gradient, is capable of "eliminating" the exterior charge aspect of a proton (via its' wake). The wake "stirs up" the FOS gradient so effectively that the neutron almost appears to have no gradient beyond the orbit of the pion/nuclear electron. Or at least none that an electron is attracted to or a proton is repelled from. If anything, a proton might actually be attracted to the space around a neutron. And in fact, has been seen in experiments. See Fundamentals In Nuclear Physics by Jean-Louis Basdevant, James Rich and Michel Spiro - a negative charge profile around neutrons. See figure 37 on the next page. This absence of an apparent FOS gradient permits the neutron to physically encounter a proton (or vice-versa). Whereas in the case of two protons their gradients, or compaction of the FOS between them, increases the fabric pressure here and thus forces them apart. By allowing a neutronic system and a proton to approach one another, the electron of the neutronic system is given a chance to be conducted from one proton to the other. And thus, force the proton it was just with to basically follow its' wake towards the other proton. If the velocity of the encounter is too high though then the electron will not have the chance to move from one proton to the other, and initiate uniting them through the power of its wake. The power of the wake being its ability to lower the FOS pressure between protons. This is supported by observation and appears to be the main factor as to why slow-moving neutrons are far more likely to be absorbed during an encounter with a nucleus than are fast ones.

---

far as its' potential wavelength or its' "re-distribution" around the proton perhaps giving the proton additional momentum/gradient for a time. Affecting the potential electron orbitals?

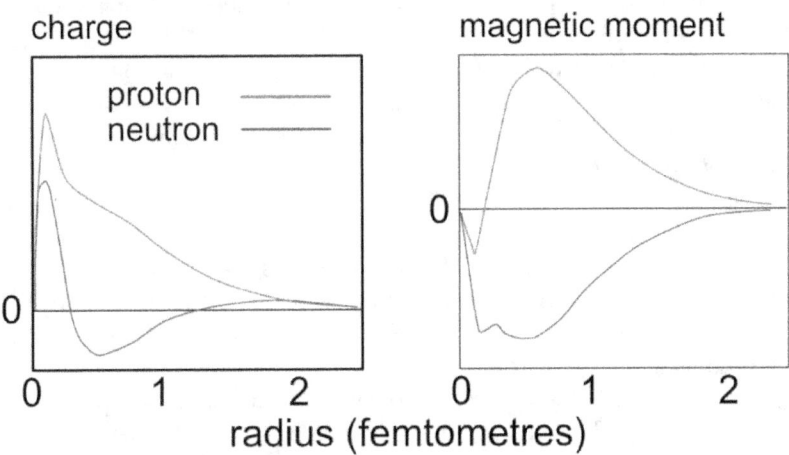

Figure 50- a negative charge profile around neutrons
[Credit: Fundamentals In Nuclear Physics by Jean-Louis Basdevant, James Rich and Michel Spiro]

The nucleus of deuterium (a proton & a neutron) is stable, unlike a lone neutron. This is made possible because during the electron's transition from one proton to the other, the gradient of the one-time neutron begins to resume its normal proton form. As the gradient begins to reform the resulting region between the two protons becomes an even stronger FOS density gradient. Which is an even more ideal conductive region for the electron (pion), as opposed to the fabric of a single proton's FOS gradient. Simultaneously this occurrence results in the protons beginning to force each other away from one another. This reaction between them is initiated by the greater concentration of aether, or density, brought about by the compression waves* between them. Thereby forcing more of the fabric into being between them and, giving rise to an even greater FOS gradient pressure between the protons (see 4.7b). This potential force that could lead to separation is offset by the electron's attraction and passage back through this region, or at least somewhere in a common plane between them, bringing the two protons back towards the same point in space once again. The force that draws the protons back towards a common point is caused by the pionic electron decreasing the fabric gradient pressure here by means of its wake. The drop in pressure makes the FOS gradient pressure on the opposing sides of the protons greater than that of the pressure of the fabric between them. The electron may well then go back to the outer side of one of the protons if it is necessary for the FOS between them to have time to become more attractive to it again. And in all likelihood is necessary for the smooth operation of the dynamics of this system. Thus, with two protons this system gives rise to an even more conductive region of FOS that alone neither of them can produce to keep the electron indefinitely in close proximity to them. Resulting in the protons sharing the electron between them.

The introduction of another neutron to deuterium, forming tritium, does not produce a stable nucleonic system (atom). After a half-life of approximately 12.4 years, an atom of tritium will decay by releasing an electron. This emission of an electron (beta particle) results in this case in the formation of a helium nucleus of an atomic mass of three. This decay of the tritium nucleus is brought about by the competition between the electrons for conduction in the increasingly rigid & dense FOS forming between the protons. With a structurally fatal accident occurring roughly every 12.4 years (thus is born tritium's half-life). The nature of this accident is a collision or near-collision between the electrons which results, in this case, in the ejection of one of them out of the nucleus. Thus, is the nature and basis of radioactivity - the decay of structurally imbalanced nuclei.

Some 453 or 31.9% of some 1420 various radioactive isotopes of the first 104 elements, that I looked at, emit electrons in a process which may result in the nuclei achieving a structurally stable form. Another 6.5% or 92 isotopes may either decay through the emission of electrons (beta particles) or by another radioactive process. Such as positron emission, orbital electron capture (E.C.(absorption of an electron from the "k" shell)), isomeric transition (I.T.), neutron emission, alpha particle emission, proton emission or spontaneous fission. Accompanying such processes can also be the release of x-rays, gamma-rays and the emission of neutrinos. Although they all sound so different, all of these nuclei are in fact unstable for the same reason. Which is that they have too few electrons, too many electrons or in the case of nuclear isomerism their electrons possess too much energy. Stated another way - these isotopes are imbalanced in their proton-neutron ratios. Too many protons (not enough electrons to hold the nuclei together) or too many "neutrons" (too many electrons competing for conductive regions between the protons). The structure of the nucleus, in terms of its stability, is based upon more than just the nucleonic bodies and the regions between them. What I'm referring to is a portion of the FOS gradient, which is commonly referred to as the Nuclear Potential Barrier. A region that extends beyond the nuclear radius a distance perhaps some multiple number of nuclear radii. The effect of this region is to act as a coulombic (charge) barrier that must be passed through by the nucleonic bodies, and the electrons, if they are to be ejected. It acts as a "gravitational well" to extend the lives of unstable isotopes. Whose nuclear resonations periodically eject nucleonic bodies (mainly alpha particles?) away from the main nucleus. And although these nucleonic bodies should be propelled away by their own positive charge once they become separated from the main nucleus the electrons that they possess are too often too strongly drawn back to the nucleus thereby bringing the protons back with them by means of their wakes. This "barrier", which in reality is only the border of a nuclear electron's domain in which it's still a force to deal with, also helps to keep stable nuclei together, and extend the lives of unstable ones. When this barrier is breached by another nuclei's barrier, but we're not talking about nuclear fusion, the electrons contained within these domains can for a short time hold these two nuclei close together to form what is known as a nuclear molecule. This nuclear molecule then exhibits a FOS gradient equivalent to their combined nuclear masses. As is shown by the resulting nuclear Zeeman energy level spectra.[Why?]

I have been considering that this barrier may actually be defined as well by the border, which beyond the wakes from the nucleus, are of a great enough disturbance to repel electrons away from the nucleus. While perhaps within the confines of this region, the wakes are not of a great enough disturbance to displace electrons from their intended paths. The border beyond this nuclear-potential barrier being the beginning of the main body of the nuclear electron wake peak expansion & dissipation region. Where the wakes are of a great enough volume/disturbance to affect an electron's path as well as being such a drop in density that the electrons that enter here, from the nucleus, may be forced to undergo a structural change once they enter into this lower potential conductibility region (weaker gradient). A change that once taken on prevents an electron from maintaining such an energetic orbit. Because now its' momentum is greater than its' conductibility, and thus the electron is carried off (ejected). The evidence being the emission of electrons from this nuclear region when nuclei possess too many neutrons (equivalent to too many electrons). But such electron emission is more likely to occur because of near-collisions between electrons in their competition for limited reasonably conductive FOS regions between protons, and thus periodically they lose their grasp on the protons (nucleons) near them.

The nucleus is quite an active body constantly resonating as the distances between the nucleonic bodies fluctuate, and in many cases, the nucleus itself rotates (Referred to as spin? (the angular momentum)).

As I stated earlier that the competition between the electrons for the spaces between the protons is like a juggling act with the forces of repulsion arising between the protons when the electrons are absent, or too far offside, bringing about the beginning of their separations, if ever so briefly. In the process, a potential proof here, causing the nuclei to resonate. Do any other theories predict the resonation of nuclei? I do not know.

Present theories cannot account for a few other items that the FOS model series can. Current theories have been able to calculate the ionization potentials of the 1s orbital electron for the various elements with a surprising degree accuracy, but fail to state why a variation in the atomic mass can have an effect upon the ionization potentials of the electrons within atoms. These potentials are only supposed to be based upon the number of protons contained within a nucleus regardless of the number of "neutrons" present. Whereas in the FOS series the ionization potential is based upon the surface area of each proton exposed to the "outside" and how the gradient of that surface area increases the density of the fabric by uniting its effect upon the FOS available to it with that of the other protons nearby. The density of the FOS surrounding the nucleus being based upon the surface area of the nucleus and the electron sharing between the protons. Add another proton and the electrons within the nucleus are forced to spend time around it - which lessens their rounds about the other protons and thus exposes more of the protons' surface areas for greater periods of time. Permitting more of the protons' wave transmissions to get through and thus compress more fabric. As opposed to the decompressive effect that the wakes of electrons have upon the fabric, making up a gradient.

The removal of an electron from a nuclear system forces other electrons to take over what FOS it had been decompressing or "stirring up". Add an electron, and this adds to the overall FOS decompression/displacement, thus permitting each electron to spend a little more time around each proton resulting in a lowering of the overall FOS gradient of the nucleus. With the surface protons requiring the least "care" by the electrons. Thus, sort of predicting a greater number of "protons" resulting in an increase in the nuclear charge at and near the surface of nuclei. Oddly enough, although not that surprising, the number of protons near the surface of the nucleus roughly approximates the atomic number of the element in question. The FOS series does not base the electron orbital characteristics (number of electrons and assigned shells) on the atomic number of the element, and in fact, my interpretation of the model predicts that the ionization potentials of the electrons are only averages. Due to the fact that the FOS density gradient of any nucleus (except hydrogen) is based upon the exposed surface areas of the protons at or near the surface of the nucleus. The Bohr ionization potential [I.P.] equation may to a fair degree of accuracy describe the first electron potentials of all the elements, but it fails to explain why the ionization potentials increase in the manner that they do. And why there is a variation due to the atomic mass when only protons are supposed to affect the charge.

The Bohr I.P. equation contains an algebraic whole number factor that represents the atomic number of the concerned element that is then squared to give us a good approximation of the I.P. for the first electron around a nucleus. And this works out quite well, but it only compensates for the outcome of the equation, and does not tell us what factors are changing within the equation so that it is "true"(works). These other factors are supposed to be constants, although there is leniency for at least one of them from the outset. That is the mass of the electron (part of the Rydberg factor of the equation) that is known to vary with velocity. So, something is varying within this equation! And it's not just the mass of the electron. What other constants are not so constant?

Particle Physics

Particle physicists have been discovering the different ways in which they can break up particles (most often protons & electrons) and accidentally create others (nearly always unstable). They give the pieces names and believe that each is an elementary particle itself. Which they've determined as being unique individuals - basing their conclusions from the tracks left behind by these particles as they pass or attempt to pass through bubble chambers, and other sensor arrays. Tracks created as they pass through the readily ionized fluid within the chamber - with the particles then knocking electrons out of their orbitals about the atoms making up the fluid. This results in the production of gas bubbles as the particles attempt or succeed in passing through these types of domains. [Include more information on some of the other types of detectors.]

All they have done is to discover what is the most likely way in which they can break up or shatter stable & unstable dense bodies of FOS, and how these doomed pieces are likely to behave & disintegrate. Stable bodies (like protons) are most often broken down when enough kinetic energy is delivered to cause them, through collision, to expand, distort & alter their internal wave patterns. Thus, resulting in their decay into positrons and photons (gamma-rays).

Someone once used the analogy that what they were doing was similar to smashing apart a watch in an attempt to uncover the gears & springs that lay within the ellipsoid. But in reality, what it's more like is smashing apart an ice cube upon some very hot surface, and into other ice cubes located here, and recording the tracks the pieces make. So that later any odd rotational energies displayed by a piece of the ice can be thoroughly analyzed and hypothesized about. The pieces are then given names as if they are completely different from what they came from.

These pieces are just what I previously stated - unstable (most often) bodies of dense FOS, which more often than not revert to electrons, positrons and gamma-rays & other forms of energy. Of course, in the meantime, an unusually structured piece could behave quite strangely as it decays into less dense bodies of FOS. Potentially stable bodies do arise from time to time, but they more often than not are doomed in normal space.

Although quarks are being referred to as the building blocks of pretty much everything their existence has yet to be proven by the collection of some clear empirical data. The main problem with the quark model is that they are said to be confined within whatever particle they make up, like a proton, and can never appear outside the particles that they make up. Quark conservation appears to fail as they proceed through decay sequences, and this contradicts the Standard Model not only in terms of disproving the fundamental nature of quarks by failure of their conservation, but in questioning their very existence. Remember that their property of *color confinement* negates their ability to ever be observed. [Once the problem is eliminated by an excuse, there is no need to reflect upon it any more. – Erwin Schrödinger] Without going into the specifics of hadron jets, color force, quark-antiquark pairs, color flux tubes and gluon tubes there is, in theory, something that can, in theory, be seen, or at least considered experimentally. The reported characteristics that make the hypothesized abundance of completely independent quarks (not part of some particle), referred to as being primordial (leftover from the Big Bang), detectable is the fact that they are fractionally charged. That is either -1/3 or +2/3 the charge of an electron or a proton. The abundance of these primordial quarks is reported to be as possibly as large as the abundance of gold in the universe. With possibly the concentration of these free quarks being higher in the heavier elements, dense metals, as opposed to the lighter elements (like hydrogen). The only experimental evidence for the proof of quarks has been derived from isolating materials in a magnetic field at temperatures near absolute zero, and eliminating all the residual charge on the particle by the addition or subtraction of electrons. If any fractional charge is leftover, roughly equivalent to 1/3 or 2/3's that of a proton or an electron, then this is considered as evidence for quarks. Such results have been found, but does experimental evidence really support their conclusions? The reason I doubt them is that all most all atoms possess additional charge, and this is what allows orbitals to be filled or shared. This is what chemistry is about. [Need to look at their experiments more.] They have come up with an excuse of why quarks are likely never to be observed directly and they call this excuse, or theory, color confinement, which is why quarks are supposed to be never directly observable or found in isolation outside a hadron, like baryons and mesons. Color confinement, from QCD, is stated as being the phenomenon that color charged particles, like gluons and quarks, cannot be separated from each other, and therefore cannot be directly observed by any experiment involving temperatures below the Hagedorn threshold of approximately 2 trillion Kelvin degrees. As Erwin Schrödinger has said: "Once the problem is eliminated by an excuse, there is no need to reflect upon it anymore."

Protons and neutrons are baryons, which are supposed to be made up of three quarks each. The proton is supposed to be made up of two up quarks and one down quark. Two up quarks have a positive charge of 2/3's each while the down quark has a negative charge of 1/3. Thus, they add up to a charge of one. While the neutron is supposed to be made up of one up quark and two down quarks. Adding up these charges produces a net charge of zero. Other quarks exist, and the various kinds can be used to build other particles. Which of course we don't ever see in natural environments. The mesons consisting not of three quarks but instead only two. An example of them being the pi-meson, or pion, which are the proposed carriers of the strong nuclear force by the exchange of virtual photons. This is where at least the current model and the FOS model concur but not by the exchange of virtual photons. This exchange was once described as the passing of a basketball harder and harder as two bodies approach one another. A very poor analogy at best which would make you think they would drive each other apart. Not bring them together. In the FOS model, it is the drop in pressure created by being in the immediate vicinity of the nuclear electron's wake that creates the effect of a strong nuclear force.

Whereas, an electron's wake in the outer regions of an atom is a weaker indirect pressure drop.

Quarks which are supposed to be fundamental are not conserved in the decay process of charged pions into muons and then into electrons or positrons. The charged pions are supposed to be composed of two quarks each, but both the muon and electron are considered elementary particles that are not composed of quarks. Worse yet for the Standard Model in an antimatter reaction with protons, the protons decay into a single positively charged pion then finally into a positron. Three quarks turns to two quarks turns into No quarks. During the process only gamma-rays and neutrinos are lost, but these are not composed of quarks either. So, quark conservation fails. The complexity of the quark model also does not help. Quantum chromodynamics [QCD] introduces another epicycle in the standard model.

It is an interesting point with QCD that is never normally raised is the implication of what it means to need to have three additional sub-types of up, and down, quarks. Which have been given the simple names of red, green and blue [quantum chromodynamics]. With the development of the Standard Model of particle physics, due to Pauli's exclusion principle, they realized the quark model required there to be three types of up and down quarks. Since no two identical particles can occupy the same quantum state within a quantum system. Some, refer to this as color charge. This means that there are in fact, by their model, nine variations on the proton and nine variations on the neutron. Definitely sounds like adding epicycles to epicycles. Add to this low energy Electron Capture, and it's interesting to note that their theory must also account for how an electron changes one of the proton's Up quarks into a Down quark to convert it into a neutron, and it always correctly adjusts, or maintains, the color somehow. Unlikely. In fact, we have the added problem of somehow a fully charged electron converts or adjusts the quark into a fractionally charged body. Then disappears. In violation of a few principles.

Ups [+2/3 charge] [red, green, blue] Downs [-1/3 charge] [red, green, blue]

Nine versions of protons formed from different quarks [2 ups, 1 down]

Up quark	Up quark	Down quark
Red	Green	Red
Red	Blue	Red
Green	Blue	Red
Red	Green	Green
Red	Blue	Green
Green	Blue	Green
Red	Green	Blue
Red	Blue	Blue
Green	Blue	Blue

Table 12 - Nine versions of protons formed from different quarks [2 ups, 1 down]

Nine versions of neutrons formed from different quarks [2 downs, 1 up]

Down quark	Down quark	Up quark
Red	Green	Red
Red	Blue	Red
Green	Blue	Red
Red	Green	Green
Red	Blue	Green
Green	Blue	Green
Red	Green	Blue
Red	Blue	Blue
Green	Blue	Blue

Table 13 - Nine versions of neutrons formed from different quarks [2 downs, 1 up]

Evidence for quarks has, in theory finally been found in the form of fractional charges. The fractional charges discovered are the result of slight variations in the FOS gradients produced by the nuclei of atoms or groups of atoms. We have already discussed the fact that an electron's negative charge does not completely eliminate a proton's positive charge (except in the case of neutrons) otherwise a second electron would not be drawn to a hydrogen atom. If it were otherwise, life would not exist. Nor would most other chemical reactions be possible. So, we would not exist. It is the FOS density gradients of nuclei that is positive charge. And any region that is dense enough for long enough shall be capable of, at least temporarily, attracting an electron. [More on fractional charges to come.]

[More on this in the future second edition of the book.]

Antimatter Problem

One of the problems that arose between cosmologist and particle physicists is that during the development of the big bang model that the particle physicists realized that both antimatter and matter should have been formed in equal amounts during the big bang. So where is it? Under their model, in some versions, there should be areas made up of antimatter. Such that there would be antimatter stars with antimatter planets. With some believing that the types of matter have somehow separated from one another. While others have proposed that both were created in the big bang in almost equal proportions but most of it was annihilated and all that remains now is just what we call normal matter. Some of this has arisen due to the balance of charge in the formation of electron-positron pairs in two gamma-ray physics, and in the generation of antimatter in high energy collisions. Consider the formation of antimatter in the high energy collisions of particle accelerators. Within the FOS model series, the formation of antiprotons they are simply a positron core surrounded by two nuclear electrons aka pionic electrons. Proof – just look at the outcome of antiproton reactions with protons and neutrons. They all indicate that in the end, antiprotons decay into one positive pion and two negative pions. If you are tearing apart nuclei, does it not make more sense that sometimes a proton is separated from the torn-apart nuclear core with an extra negative pion? This seems far more likely than the alternative of bringing into existence of a "new" type of matter not found in our solar system. Unbalanced nuclear wreckage is for more likely an explanation and is supported by the decay products. Based on the experiments of Hans Mes and Jacques Hébert.

Nuclear Density and The Neutrino

I suspect that the actual distance at which an electron revolves about a proton, in the formation of a "neutron" is on the order of somewhere around one to ten times (?) the diameter of the proton. It is theoretically structurally important to determine at least an approximate value - for the resulting intra-nuclear distances between protons will determine the likelihood of how far a neutrino may travel through a given amount of material before it is stopped by an encounter with a nucleus. In other words - Just how solid is a nucleus? If the distance is on the order of say something like ten times the diameter of a proton, then this would make it possible for some neutrinos to pass through the smaller nuclei with relative ease. However, the larger the nuclei the less likely they would be able to pass through a similar amount of material for the reason that the number of protons potentially blocking their paths by the formation of a "wall" would be significantly greater if the protons were more greatly separated. But of course, there is a certain limit if surpassed the neutrino would then see a large nucleus like we might see a flock of birds. So, spread out that few things would have a problem passing through them. Since it would then be composed mostly of empty space.

The ways in which electrons and neutrinos interact also pose a physical structure for neutrinos. In all likelihood neutrinos are an odd part of the electromagnetic spectrum. Unlike gamma-rays which are from relatively large volumetric changes in the aether gradient neutrinos seem to be related to a high energy wave type but of a relatively small volume of high density aether as compared to the physical volume of gamma-rays. Is this all that they really are? Since they interact so rarely with nuclei, and electrons, this would seem to be the most simplistic and thus logical explanation.

[Insert new material on the way neutrinos are ejected material below the normal photon-type requirements. Or as excess liberated material lost by nuclear-electron interactions. In other words, they are a smaller than normal type of photon wave.]

[More on this in the future second edition of the book.]

The Anti-Proton and Pions

I use the sentence; "It is believed that all particles have an anti-particle equivalent.", because not all of the so-called anti-particles are necessarily what they are believed to be. Perhaps, more importantly, is that the majority quickly decay into the more fundamental stable particles – electrons, positrons, neutrinos and gamma-rays. Some of the best empirical data that reveals the nature of matter and anti-matter come from Hans Mes and Jacques Hebert from their study of annihilation reactions into pions. As well as the complimentary data from antiproton-neutron annihilation studies by Mario Gaspero. Note that I am referring to his experiments where antiprotons are reacting with neutrons to annihilate each other.

It has been reported that when an anti-proton strikes a proton, the proton becomes a neutron while the anti-proton is transformed into an anti-neutron. This anti-neutron is the same as a neutron except for the fact that it decays via what is termed "star fragmentation". Actually, the experiment performed by Chamberlain and Segre was performed to observe anti-neutrons, but they utilized anti-protons to bring into existence anti-neutrons. The following is reportedly the sequence of events and what observations they were interpreted from: "the anti-neutron can be detected... [Content removed until further funding acquired.]

By definition, an anti-neutron is a particle with the same mass, but having a reversed magnetic moment of equal magnitude. In other words, the magnetic moment of a neutron is parallel to the angular momentum of the "neutron" while in the anti-neutron its magnetic moment is anti-parallel to its angular momentum. There is another interpretation of the events in the bubble chamber that also fit the tracks left behind by the passage of the particle. This is not at present an accepted interpretation of the data but it would fit in well with the FOS model.

The curved track that is supposed to be an antiproton, is in fact, the track of a newly released particle system of two electrons and a proton - "a charged neutron". Permitted to exist for a short time while the electrons attempt to secure an exclusive hold onto the proton while in direct conflict with one another. This "charged-neutron", or anti-proton as it is normally called, then encounters a proton to which the nearest electron finds itself conducted to. Thus, converting some proton into a neutron, and the former anti-proton (charged neutron) into a "disturbed" neutron that then no longer leaves a trail in the bubble chamber because it is neutral. But in its ejection from some nucleus, too much energy was delivered to the proton and it has become so unstable that it breaks up into smaller pieces of high-density FOS. With it then breaking up into five pions, which they in turn themselves decay. I wonder if the existence of the "charged" neutron may have been helped by the possibility that its proton was no longer stable and beginning to expand and break up. Thus, changing its FOS gradient so that another electron is permitted to orbit it for a brief, but longer than usual amount of time. Although it's something to consider I don't think that this is the usual case for the existence of anti-protons (charged neutrons).

Recently [1992] I finally got around to searching for information on anti-matter and anti-atoms. I was both surprised and relieved to discover that anti-atoms are not a common laboratory resident. And in fact, into the year 1991 stationary antiprotons have not been kept in existence beyond some 100 seconds. Let alone being used to form anti-atoms. I was surprised because the literature gives the impression that anti-atoms are not only a common occurrence, but as stable as any other normal atom. First of all, no anti-atoms have been formed within the laboratory as of 1992. [Not entirely true anymore. See antihydrogen.] Secondly, if you know how antiprotons and antineutrons are generally formed, via the collisions between high-velocity nuclei with other nuclei, does it not seem more likely that the collision products might contain a nucleon with two quanta of nuclear stabilizing particles of the "strong force" (in other words two nuclear electrons) that had been dislodged from the nucleus during the collision and simply took up residence around the only proton available to them at the time. Or does it seem more likely that an alien particle from an anti-universe has broken through space to find itself in our universe? Brought here by breaking a hole in the fabric of space & time across multi-dimensional channels. A cheap shot? A more reasonable theory would simply have a particle system, induced by an extremely energetic collision, formed by chance in which two pion-electrons (nuclear electrons) are merely competing for the domination of a proton's FOS gradient. In the collision that created this situation did these particles become "damaged" such that their normal structural characteristic values are different? Resulting in somewhat different gradient generation and deformation values allowing two pionic electrons to share a damaged proton/positron? Since these two nuclear electrons are competing for the gradient of the lone proton, they are unable to confine their attraction to it and are therefore attracted to other FOS density gradients. Thus, this heavy particle system is termed negatively charged because it is attracted to other protons and nuclei. The best part of all though is that this proposes not only a new experiment for a proof of at least a portion of the FOS model series, but also more importantly, one that is already being actively pursued. That experiment being the formation and continued existence of antiprotons and anti-atoms.

Usually, a lone electron will isolate a proton from the "rest of the universe" within the particle system we call a neutron. This system decays if it does not enter a nucleus and share this nuclear electron with another proton. If it cannot find another proton then some 600+ seconds later the neutron will break apart. As I stated earlier - it should not be surprising to find that during these extremely energetic collisions that sometimes when these nuclei are shattered that occasionally a proton is dislodged with two pions (nuclear electrons) competing for "dominance" of the FOS gradient of this proton. Competing about one of the protons they had just, a short time ago, been readily sharing. They are just sharing because they are also traveling around other protons as well. Since the pions (nuclear electrons) are competing for the fabric of the gradient about the proton, they are not as strongly attached to it, and so are attracted to other FOS gradients. Either full gradients or pockets of denser material. As a result of such orbital movements an anti-proton should possess a greater wavelength than neutrons of similar energies or velocities (?). More importantly though the FOS model predicts that these antiprotons should be more unstable than non-nuclear neutrons who have a half-life of less than 11 minutes. The 100+ seconds that is the limit of their stationary storage times, so far, of anti-protons may in fact be their half-life.

I deliberately used the word stationary since by moving an electron at the high velocities of accelerator storage rings the masses of the anti-protons are much higher than normal. And since mass and positive charge are the very same thing, this means that a high-speed anti-proton will possess a greater FOS gradient and therefore be able to hold onto an electron for longer periods of time. In fact, once its mass is equivalent to approximately the gradient between two protons, like the situation within a nucleus, then any such body should be able to hold onto an electron indefinitely. I had at an earlier time failed to realize that an accelerating anti-proton's mass would be dramatically increased within an accelerator storage ring, and since positive charge & mass are one and the same, in essence, then so would the attraction of any nuclear electrons to them. I had forgotten about the implication of motion induced gradient formation (which is mass) that is until someone pointed out to me that anti-protons last for days within storage rings. And if I was considering this as one of my major proofs that I had better rethink my theory. I knew immediately that I had forgotten to clarify that this half-life of 100+ seconds was only for stationary anti-protons. In fact in my original paper that I had submitted to this person that in the following paragraph of commenting on this storage time as a possible proof for my theory I had stated that:- "It may be possible to extend the lives of these rare particle systems beyond their present half-lives by keeping them stored at much higher velocities". From that same paragraph:-"by forcing them to move at greater velocities through the surrounding fabric of space - combined with their spins permits the parent protons to possess greater FOS gradients, by compression, about themselves and thus attract electrons more strongly & thus over longer periods of time."

In December of 1992 [source of information missing], I was informed about the fact the ultra-cold anti-protons can be kept in storage for months. At first, I was not sure what to think about it until I read the article in Scientific American. I hate having to account for phenomenon that I had not conceived of but I still can account for this phenomenon without violating any of the FOS Theory concepts. Because after studying the article I realized they had accomplished what I thought could only be done by increasing the mass of the particle by doing simply the reverse. That is making the surrounding FOS in the neighborhood of the antiproton "hostile" or at least a very bad conductor to electrons as opposed to the gradient about a single proton. Thus, making the weak gradient a much more favorable conductor than it would normally be than the fabric of space outside the particle system. I also think that the forced magnetic alignment by the confinement coil must also help with preventing the anti-protons from decaying more rapidly. It still stands that under normal fabric conditions the half-life of anti-protons is less than that of neutrons. Something else that then came to mind was could this technology be used to extend the lives of ultra-cold neutrons?

Positrons that for a time might form some sort of a system with an antiproton but should not form the anti-hydrogen atom that so many have assumed exists somewhere. In fact, according to my interpretation of the FOS Theory series, no anti-atoms of any type should be able to form naturally. Thus, no anti-atom spectral signatures from anti-atom stars. The formation of this so-called potential anti-hydrogen atom should in fact accelerate the destruction or decay of the anti-proton. The positron being annihilated by one of the nuclear electrons (pions) orbiting the proton. Perhaps resulting in the emission of two gamma-rays (or two neutral pions) and a neutron or anti-neutron from the confinement vessel.

The FOS series can also easily explain why sometimes in the collision events between two protons an electron-pion and a positron-pion are formed. Here we have an event with similar consequences as the collision between two gamma-rays except that now the FOS is in residence about the surface of the protons, not of ordinary space, and thus already of comparable density &

rigidity of gamma-rays. The likely event of the formation being the collision, induced from the spin of the protons, between opposing waves within the two gradients. Resulting in one wave passing through some slightly less dense & rigid fabric. Almost exactly like the events in the formation of an electron-positron pair from the collision between two gamma-rays. Another reaction with similar consequences that can be explained basically the same way is the formation of an electron-positron pair from the collision between two electrons, probably nuclear, this time being formed from the collision of the fabric they are carrying in their cone-like depressions. If they had just left nuclei, then the density & rigidity of the fabric they'd be carrying would be comparable to that of gamma-rays.

[More on Hans Mes's work et al.]

Anti-protons decay, on average, into two electrons and one positron – starting out as two negative pions and one positive pion at first. Under annihilation reactions with anti-protons – neutrons decay into a single negative pion and one positive pion then finally as they adapt to regular density space and shed "mass" into an electron and positron. The lower energy decay of a neutron producing just an electron and a proton. Here the positron stays as a proton since its accumulated aether gradient [mass] is retained. Most particles decay into some combination of pions but in the end into just electrons, positrons, neutrinos and gamma-rays. The standard model should be called the standard model of high energy states of the fundamental particles. In the end a number of their particles are irrelevant, because they are not fundamental. Especially the color variants of quantum chromodynamics which imply nine types of protons and nine types of neutrons.

Papers

Antiproton Annihilation And Pion Interactions In Complex Nuclei by Hans Mes [1968]
Document ID NK04743

Antiproton Proton Annihilation Into Pions
Jacques Hebert and Hans Mes [1967]

Study of a Narrow pi+ pi- Peak at about 755 MeV/c^2 in antiproton-neutron -> 2pi+ 3pi- Annihilation at Rest
by Mario Gaspero
Document ID 1005.2381

The Standard Model versus The Fundamental Model

The Standard Model is supposed to be the culmination of our civilizations' understanding of the fundamental particles, and forms of energy, that make up the observable universe. It mainly consists of three groups of particles; quarks which give us protons and neutrons, leptons which consist of electrons and their relatives and their associated neutrinos, and the force carriers of which photons are part of it. And place holders for the Higgs particle and Graviton. The quarks and leptons are grouped together as the constituents of matter, and this group is classified as fermions, while the force carriers are grouped together under the classification of bosons. The fermions are also categorized into three "generations of matter." These generations are more accurately described as energy states of the first generation or family. The tau decays into muons or electrons, while the muons decay into electrons. So, in the end you end up with just electrons. And the same is true for the quarks with the final products being up and down quarks.

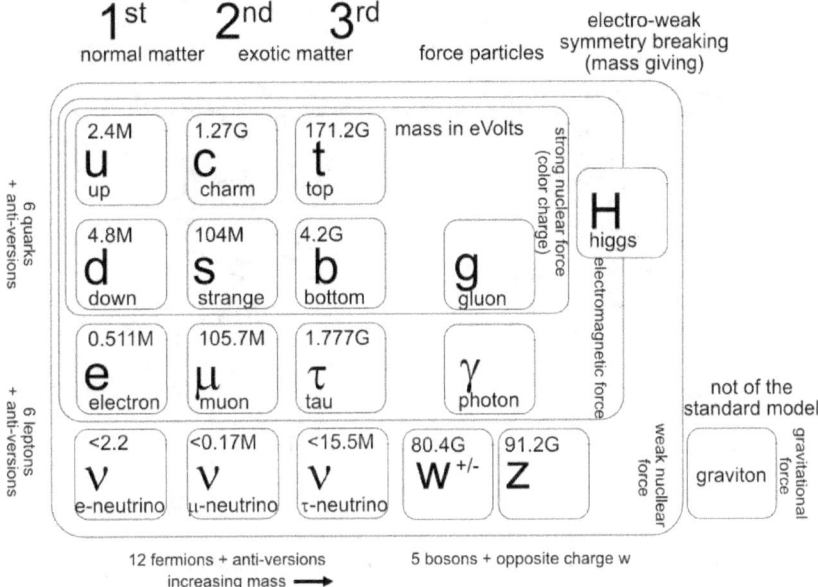

Figure 51- The Standard model

The Death of the Dark Energy Idea

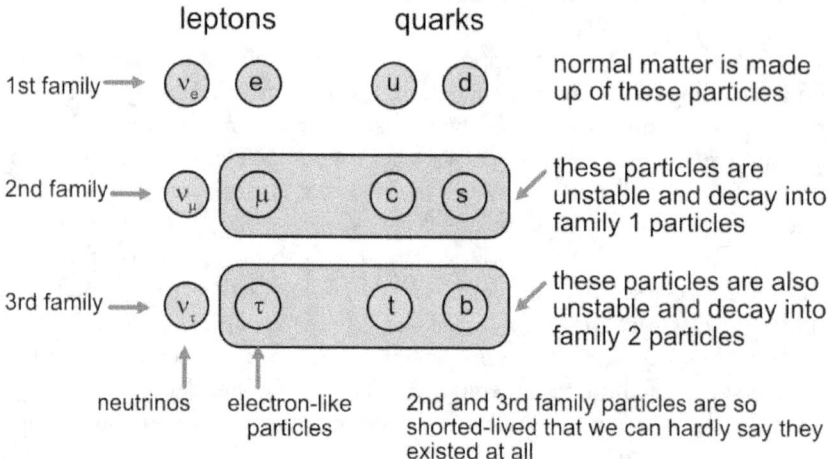

Figure 52- The Standard model families or hierarchy

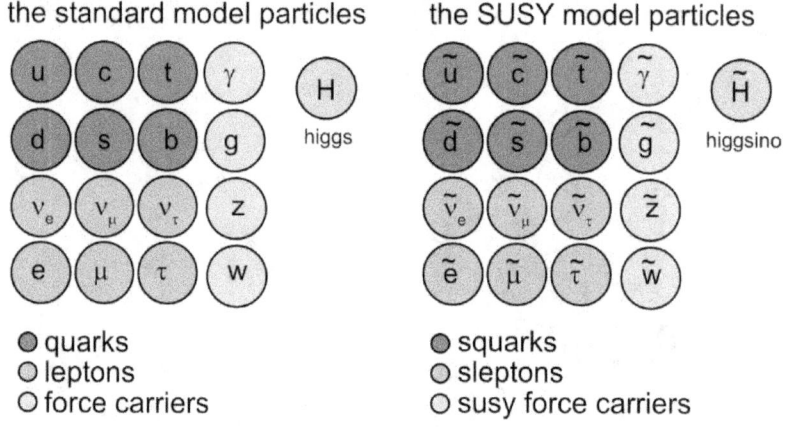

Figure 53- The Standard model with SUSY, supersymmetry and sparticles

The experiments, using a variety of different particle accelerators, are based upon the idea that if you add more energy to the collision of particles, it will reveal particles that were hidden from view. But if you look at what these particles quickly decay into it is these final particles that seem to be the true building blocks of everything. Electrons, positrons, gamma-rays [et al.], and neutrinos. The standard model is dominated by a collection of unstable man-made nuclear wreckage and a few hypothetical particles needed to fill out a table to cover all the supposed forces they've envisioned.

The Fundamental Model, proposed in this book, of particle physics is a collection of only the stable particles that still remain after all the high energy exotic particles have shed their excess energy, and mass, and left behind only the electrons, positrons, photons [waves from the electromagnetic spectrum] and neutrinos.

The fundamental model does not contain quarks, gluons, pions, muons, gravitons, ... they are not necessary and only exist as either mathematical oddities or high energy versions of the fundamental particles. The first table of the original Standard Model is shown with the parts that are strictly high energy version eliminated and the unnecessary quarks eliminated.

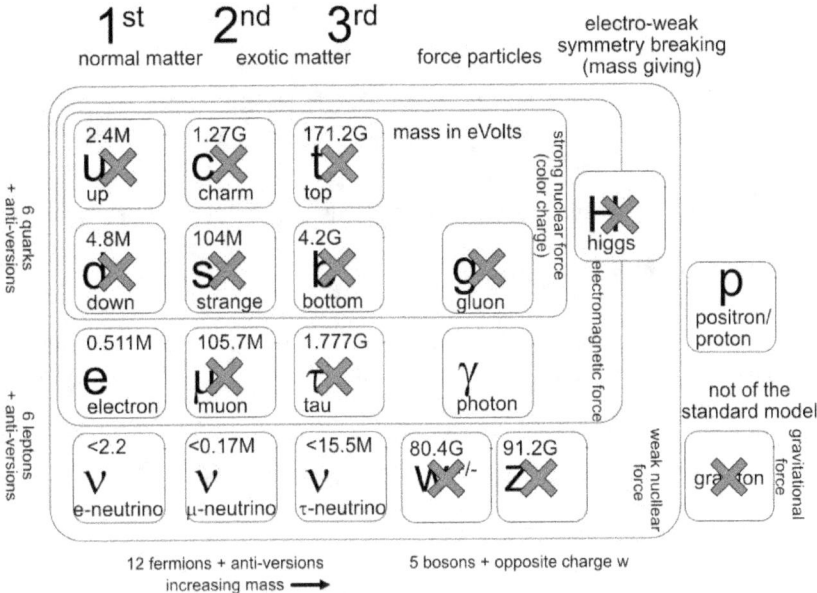

Figure 54- *The refined Standard model or Fundamental model*

The new table shows the Fundamental Model required for the new FOS aether model to account for the observable universe. With just one addition of the positron/proton. And 12 subtractions. The neutrinos remain but should be adjusted to reflect their re-classification as a type of volumetrically-smaller than normal high-energy photon.

Chapter Five

MAGNETISM

The greatest obstacle to discovery is not ignorance—it is the illusion of knowledge.
– Daniel J. Boorstin (1914-2004)

The greatest enemy of knowledge is not ignorance, it is the illusion of knowledge.
– Stephen W. Hawking (1942-2018)

The surest way to corrupt a youth is to instruct him to hold in higher esteem those who think alike than those who think differently.
– Friedrich W. Nietzsche (1844 - 1900)

Magnetism

Of all the forces in nature, magnetism has to be the most fascinating of all for the simple reason that its effects can so readily be observed and manipulated. The rise and suspension of iron filings appear to us as almost magical. The repulsion, or attraction, of two bar magnets while not affecting most of the matter around them is fascinating to all of us. Superconductors floating over magnets astound us.

One thing we did not consider when we looked at, and hypothesized about the structure of an electron, was whether or not it possessed rotational energy. Rotating clockwise or counterclockwise depending on whether or not it is approaching an observer, or moving away from them. Well, it appears they do so. Experimental proof of the hypothesis that light possess angular momentum was provided first by Richard A. Beth in 1936. [A picture of this would be good and a note in the glossary.] He "showed that when circularly polarized light is produced in a doubly refracting slab, the slab experiences a reaction torque."

The angular momentum of photons plays an important role in both classical & quantum mechanics. For without it the angular momentum of atoms would not be conserved when photons are emitted. In James Clerk Maxwell's book – A Dynamical Theory of the Electromagnetic Field – he introduces us to not only to the idea of electromagnetic momentum of currents, but also of the magnetic field. Beth's work happening decades after Maxwell's passing [born 13 June 1831 – died 5 November 1879]. If had he had this critical piece of evidence, this would likely have been one of the most important pieces of information that could have allowed him to take his theory even further.

It should not be surprising then that when a gamma-ray undergoes the transformation into becoming an electron that this rotational energy not only becomes more apparent, but is also further developed into becoming a prominent feature of the electron's structure and its effect on the FOS it passes through. The reason we're concerned about whether or not an electron rotates, and the rate at which it does so, is that this appears to be a good candidate for the structural component that gives us magnetism. Electron rotation was proposed in 1925, and long ago accepted as fact, to account for additional components in the spectral lines of matter thought to be revealed through and as a result of the Zeeman Effect.[More on the Zeeman Effect?] How can such rotation give us magnetism and the physical effects that we observe? [Also, what about the movement of the pilot wave material as it moves out of the way of the passing photon? Its motion would be perpendicular to the electric field. Does this motion contribute to the magnetic field being generated, or is it strictly angular momentum?]

We now return to the equation linking electromagnetism to the speed of light. This equation was seen as the proof linking photons and electromagnetism together, and which is the source from which James Clerk Maxwell derived the concept of the Electromagnetic Spectrum and predicted the existence of radio waves.

It has been found that if one divides the electrical force constant by the magnetic force constant one ends up with the velocity of light squared.

$c^2 = K_e / K_m$

The electrical force constant [from Wikipedia]
[Also called the vacuum permittivity or permittivity of free space.]
$K_e \approx 8.9875 \times 10^9$ N·m² / C² or kg·m³/s²/C²
$K_e = \dfrac{1}{4\pi\epsilon}$

The magnetic force constant [from Wikipedia]
[Also called the vacuum permeability or permeability of free space.]
$K_m \approx 1.2566370614... \times 10^{-6}$ N·s² / C² or (kg·m/s²)·s² / C² or kg·m/C²
$K_m = 4\pi \times 10^{-7}$
Other units H / m or T·m / A or Wb / (A·m) or V·s / (A·m)

Where capital C is for Coulomb. So, with respect to a coulomb of charge.

Now we introduced the hypothesis that this equation not only links photons and electromagnetism, but it also more importantly it appears to be proof for photons being of a longitudinal nature, and not transverse waves as is accepted by the experimental so-called proof of polarization. Remember that in an earlier chapter, we showed that longitudinal photons that have an ellipsoidal profile are also polarizable. The equation clearly shows that the electron is "born" from a photon (gamma-ray) and that the energy of the photon is still within the electron, but has taken on new forms [characteristics]. The constants themselves should well tell us more precisely about the quantitative characteristics of the structure of electrons and how this corresponds to the nature of electricity and magnetism. Also, circular polarization may be an indicator of the longitudinal nature of photons as opposed to their being transverse in nature. It is easy to envision triggering circular polarization in photons if it's just a mass of material, but not if they are "two-dimensional" transverse waves. What could cause and support the rotation of a transverse wave? And remember in transverse wave conduction that they are boundary wave distortions. So, in a transverse wave model for circular polarization the boundary would also have to be rotating. I can't envision this. Can you?

The standard equation for the velocity of longitudinal waves within a medium consists of the Bulk Modulus of Elasticity divided by the Density of the medium, which then provides us with the square of the velocity of the waves in that medium. For longitudinal waves, the bulk modulus is defined as the pressure increase relative to the decrease in the volume. The bulk modulus of a medium is a measure of how resistant to compression it possesses. By mathematical definition it is the ratio of the infinitesimal pressure increase to the resulting relative decrease of the volume.

How does this pertain to the speed of light? Let us look at the units of the electrical force constant:

$K_e \approx 8.9875 \times 10^9$ **N·m² / C²** or **kg·m³/s²/C²**

In particular note the **m³/s²** which seems to indicate cubic metres per seconds squared. Velocity is m/s, and acceleration is m/s², could this then imply we could be looking at a rate of change of the change in volume. That is, it could be a measure of how fast the speed in the change in the mass-volume relationship is occurring. If read directly from the units it can be seen as some measure of pressure in Newtons on an area. **N·m²**

Volumetric and pressure changes is exactly what we have defined as the true nature of electrical component of electrons. This is no coincidence, it is proof.

The second component of the standard longitudinal wave equation is density which is defined a mass per unit volume. Let's look at the units for the magnetic force constant:
$K_m \approx 1.2566370614... \times 10^{-6}$ **N·s² / C²** or **(kg·m/s²)·s² / C²** or **kg·m/C²**

The N·s can be linked to the unit of momentum. From F = dp/dt, where you can restate it as 1 N·s = 1 (kg·m/s²)·s = 1 kg·m/s [mass·velocity].

Can we then consider the energy by looking at the momentum N·s acting for one second. N·s²? Thus reducing it to the just the units of kg·m due to this action.[?]

[More on this in the future second edition of the book.]

The momentum of the aether/FOS is what we have said is related to the nature of magnetism, or alternatively the forced motion of electrons so that any gradient change acts as if momentum is acting on them. Let us look now at this in greater detail. It is the electron's sensitivity to changes in gradients and the motion of the aether that is so important to most of what we consider magnetism. Just as the path of photons are affected by the gravitational fields of stars, our slowed down "gamma-ray electron" forms are even more sensitive to all aether gradients, gravity and the electric fields, and even their changes and motion.

Links
https://en.wikipedia.org/wiki/Maxwell's_equations
https://en.wikipedia.org/wiki/James_Clerk_Maxwell

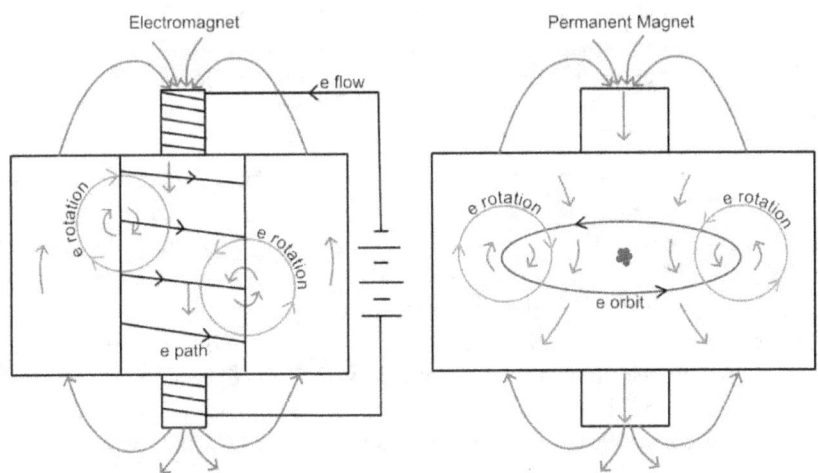

Figure 55 Electromagnet and Permanent Magnet and their electron motions.

Permanent Magnets

The electron has two physical features that have an affect on the fabric it passes through, and then there is, of course, Louis de Broglie's [and Sid Deutch's] pilot waves. At the front of the electron the material ahead of it is experiencing compression and can become an emitted photon if the electron undergoes a dramatic change in direction. A synchrotronic light source is such a device that can be used to emit photons in such a manner. In all likelihood both the photon and pilot waves gain rotational momentum from the electron's rotation. Evidence for this arising from the reaction torques measured in Beth's double refracting slabs, and perhaps the left-handed rotation of neutrinos. Which neither classical nor quantum mechanics has been able to explain. Does the Faraday effect[15] indicate that it might be possible to have some electrons formed with a right-handed twist? As a result of this inwardly rotating "ring" drawing FOS through an electron's center it seems to create something akin to a vortex in the surrounding fabric as a result of its rotational energy. Ignoring the pressure drop, created by the generated wake, for the time being, which is the negative aspect of what we call electrical charge. Combining these two actions, compression and rotation, initiates the FOS that passes through the electron into circular motion through its' center and at least adds some component of circular motion about its former path. Does some of this action simply spit out the material and eject it outward from the path of the electron? With the material simply flying off until its momentum is absorbed? And although the electron itself can appear to be vortex-like within the FOS, it is not until an electron either combines its behavior with other electrons in forming atomic magnetic cells, or they combine their activity in the formation of an electrical current along some path, that a noticeable magnetic field can be detected.

Note that during the exploration and development of this model/book that too many sources of information, in regards to electromagnetism, are often describing current flows by so-called conventional current versus electron current. For historical reasons many sources often refer to the so-called positive charge flow as current, and not electron flow, along a wire in their experiments. Too often, I was convinced they were talking about one "current" when, in reality, they were talking about the other. Worse yet some sources refer to both at the same time, but it is not always apparent that they know it. Some will denote the conventional current with a subscript on 'I' with a c: $= I_c$

---

[15] The Faraday effect was the first experimental evidence that light and electromagnetism were related.

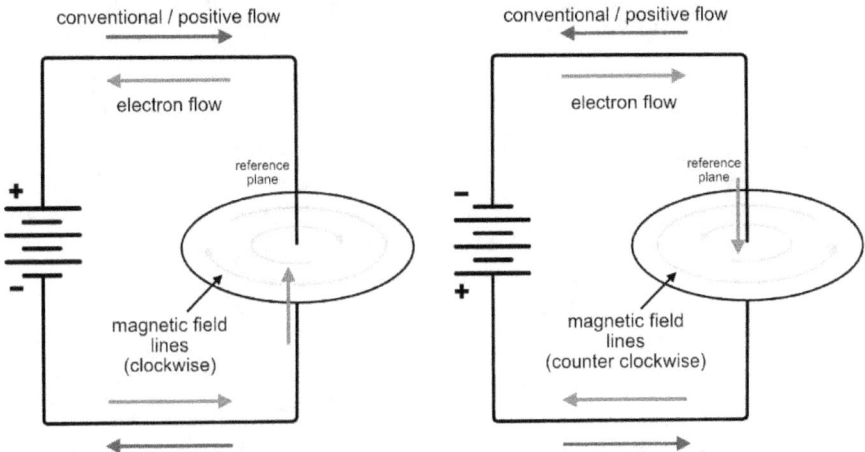

Figure 56- conventional current vs electron current

Magnetic fields have two origins:

1. For atomic magnetic cells at least one electron, or two, takes up a crudely hewn single planar orientation around a nucleus, and their collaboration with others nearby, trigger the formation of a structurally detectable vortex-like flow - a magnetic field. Not that they occupy a single plane, but the sum of their orbital motions favors a single plane. Is it pairs of electrons or is it a single orbital electron that can normally take up these single planar orientations? Normally the presence of another electron too nearby competes for the "re-normalized" FOS and this forces the pair into precession about the nucleus. Thus, potentially preventing any single planar orientation being achieved, or if their separation is enough then the FOS gradient orbital region has enough time to recover then would this still allow a pair to remain enough into roughly a single plane. Is it the action of a pair of electrons whose influence on one another's orbitals that allows them to take up a singular planar orientation? Or is it the influence of other electrons nearby in other orbitals that aid them in taking up a singular planar orbital pattern? Or both? What is the electron orbital pattern of iron and other magnetic elements? Current models of atomic magnetism often invoke looking at pairs of electrons in their models to talk about filled and partially filled shells based on the pairing of electrons. However, if one looks at the ionization energies of all the electrons in orbit around the various different chemical elements this idea is not directly obvious from the data. [Are they looking at the orbital shapes? Should we include an image at this point of the orbital shapes?]

2. For electromagnetics electrons are forced along wires to take up similar planar orientations and induce magnetic fields originating now outside the atom. Here the electrons are freer to move and interact at a different level. What

are the consequences of this for the two types of magnetism, and how might they interact or behave differently?

It is iron $^{56}_{26}Fe$, cobalt $^{59}_{27}Co$, nickel $^{59}_{28}Ni$ and some alloys of the rare earth elements, like neodymium $^{144}_{60}Nd$ and samarium $^{150}_{62}Sm$, that are used to make permanent magnets. The following graph is of the electron-volt [eV] energies of all the electrons of four elements to show how even the values of electron energies appear. For Argon, Vanadium, Iron and Copper.

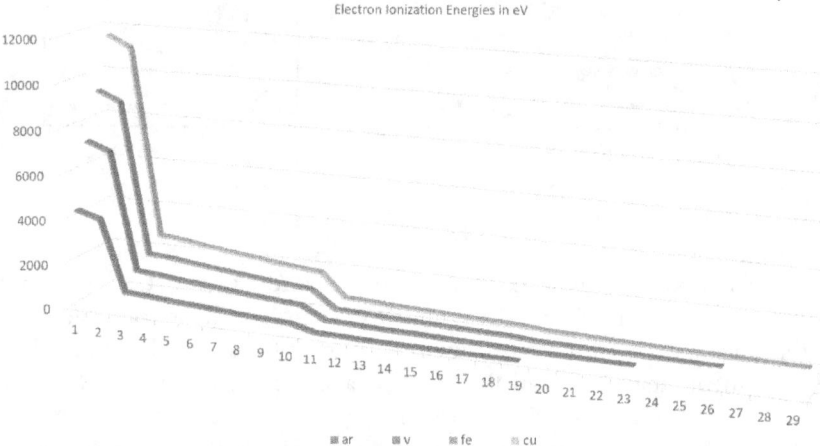

Figure 57- Ar, V, Fe and Cu ionization energies in eVolts

If we look at the standard diagrams often used to portray how the electrons take up positions (orbitals) around nuclei it helps us see what they mean by lone electrons in orbital positions, but lone electrons are common among non-magnetic elements as well. The following Aufbau diagram shows us the standard energy electrons around a Nickel nucleus. The Aufbau principle, or building up principle, is used to show how the electrons fill the lower energy orbitals first.

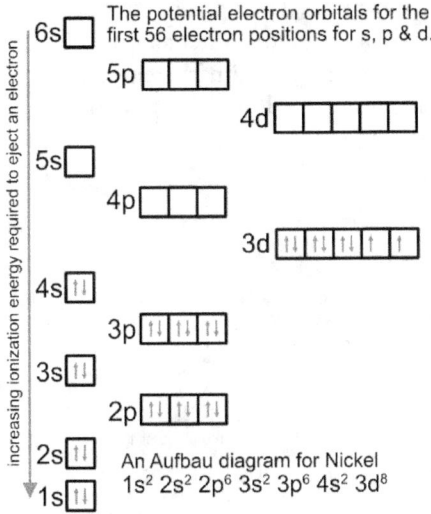

Figure 58-An Aufbau diagram for Nickel

Electron orbital energies in electron volts (eV) for Iron, Cobalt and Nickel

Fe: $1s^2\ 2s^2\ 2p^6\ 3s^2\ 3p^6\ 4s^2\ 3d^6$
Co: $1s^2\ 2s^2\ 2p^6\ 3s^2\ 3p^6\ 4s^2\ 3d^7$
Ni: $1s^2\ 2s^2\ 2p^6\ 3s^2\ 3p^6\ 4s^2\ 3d^8$

	$1s^2$		$2s^2$		$2p^6$					
Fe:	9277.69	8828.00	2023.00	1950.00	1799.00	1689.00	1582.00	1456.00	1358.00	1266.00
Co:	10012.12	9544.10	2219.00	2119.00	1962.00	1846.00	1735.00	1603.00	1504.60	1397.20
1Ni:	0775.40	10288.80	2399.20	2295.00	2131.00	2011.00	1894.00	1756.00	1648.00	1541.00

	$3s^2$		$3p^6$						
Fe:	489.26	457.00	392.20	361.00	330.80	290.20	262.10	233.60	
Co:	546.58	511.96	444.00	411.00	379.00	336.00	305.00	275.40	
Ni:	607.06	571.08	499.00	464.00	430.00	384.00	352.00	321.00	

	$4s^2$		$3d^6/3d^7/3d^8$							
Fe:	151.06	124.98	99.10	75.00	54.80	30.65	16.19	7.90		
Co:	186.13	157.80	128.90	102.00	79.50	51.30	33.50	17.08	7.88	
Ni:	224.60	193.00	162.00	133.00	108.00	76.06	54.90	35.19	18.17	7.64

Figure 59-Electron orbital energies in eV for Fe, Co and Ni from the CRC Handbook

Some additional things we need to look at regarding magnetism, to come up with a physically-based model for it, is to understand how free-to-move charged particles behave in magnetic fields, and how constrained particles behave. When we are talking about constrained here, we are talking about electrons bound to a wire, or an atom. Positively charged particles that were propelled into a magnetic field, that are not constrained in their movement, move in the opposite direction to the paths taken by negatively charged particles. See the next figure. The figure after that shows the interaction of an electron current-carrying wire between the poles of two magnets. The right-hand motor rule can be invoked to describe the second figure and the interaction between the magnetic field of the magnets and that of the electron current-carrying wire. Note that if you apply the right-hand motor rule to the first figure, of an unconstrained electron, you see that it helps describe the arcing motion of the electron about the magnetic field lines.

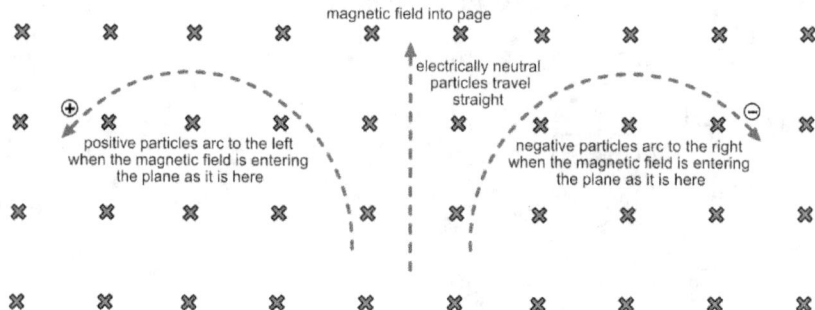

Figure 60-positive, neutral & negative particles moving thru a magnetic field

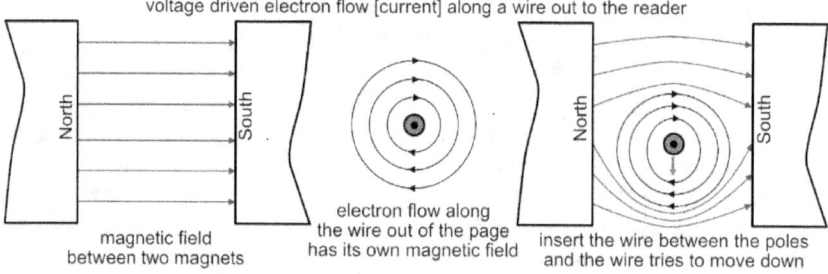

Figure 61- voltage driven electron flow along a wire between to magnets

Concept 1

Normally it is not until the vortex-like action induced by an electron is combined with many others, generated by neighboring electrons about other nuclei, that an easily detectable vortex / magnetic field is generated. Ignoring for the moment electromagnets. With each little nuclear magnet adding energy to the flow of the fabric - similar in effect to combining the efforts of more and more electric fans in an attempt to circulate a body of air. [Could we look at the mathematics of adding or multiplying the increase in the movement of the volume of the aether/FOS per unit of time or perhaps something like momentum per unit volume per unit of time? Or some kind of velocity?] The FOS is brought into motion as the result of both the compression of the fabric in front of the electron, its' densest portion of its pilot wave, and the electron's poloidal and toroidal rotations. This effect results from the transitory "adherence" of the gradient forming fabric in the general area responsible for its compression, due to the electrons conduction, causing electrons not only to move & compress the FOS even further within themselves, but also to set it into circular motion by way of the electron's rotation along its axis of travel, or what is normally referred to as spin[?]. Moving FOS seems to affect relatively slow protons, not the stationary ones, basically in the same manner as a motor boat is affected by the current of a river that they might be within. [More on this coming up.] But electrons seem to move more under the influence of the density of their conductor, just as is observed for photons that pass closely to the surface of the Sun during solar eclipses, and the electrons interact with flows only if the FOS gradient they also reside within, around atoms, gives them some freedom to interact with external magnetic fields. Generally, the strength of conduction of an electron around its' parent atomic/nuclear gradient is greater than the influence of any moving flow of the FOS/aether outside the atom. A chemical bond that is too strong provides little opportunity for an electron to be affected by other factors. An orbital bound electron is not a free electron. More often than not because of the competition between electrons, for an orbital orientation that would complement the external magnetic field, this prevents the electrons from orienting themselves to it, because of their effect on the surrounding fabric of the gradient about & within the orbitals through the action of their disruptive wakes. Thus, preventing most electrons through their disturbance of the fabric from taking up a magnetically inducing planar orbital.

Concept 2

The action of the electrons revolving about a common center triggers a higher density region due to aether particle motion towards the center of motion. This appears to form a lower density region on the outside area of influence and higher density region in the center. Particle density would increase due to the motion of more particles into a region. The opposite would also be true. A lower density region around boundary regions of magnets. The higher density aether would flow to the lower density region and give us vortex motion to some degree. Would this provide another mechanism for the action and motion of interacting magnets due to the electrons reacting to these vortex-like gradients? The formation of higher and lower density regions, due to the combined action of the electrons, might also in part play some component in the motion of two parallel current-carrying wires which we will look at.

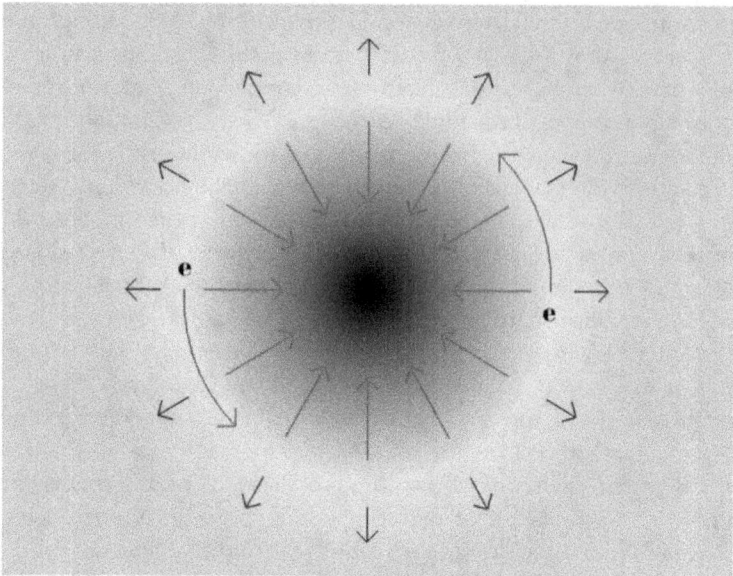

Figure 62- electron induced aether flows around a circular path

Concept 3

Having the electrons moving material and forming a denser region in some common center would not only increase density and thus provide a reason for electrons to move into these areas. But it also would create a flow more active at the center of the magnetic field, and also change the dynamics of all moving particles that enter such a region. Thus, we have changes in motion, triggered as the electron frontal-load, and their pilot waves, encounters and is struck by the moving material of the magnetic flow, triggering differential pressure and density regions. Similar to the situation that sets up conditions for kicked balls to experience the Magnus effect, but with the critical difference that the electrons are in essence a "special" type of photon being conducted by a medium and as such are particularly sensitive to density gradients and density pockets. That being considered they are being conducted through space and are influenced, and forced to align themselves, due to the surrounding material ahead of and around them with the aether/magnetic flow, such that the electrons are easily conducted into denser regions that they themselves help to form. The very opposite of what happens in the Magnus effect for non-conducting bodies. The moving proton appears to experience the standard Magnus-like effect of spinning bodies. The physics of the Magnus effect is triggered by a spinning body as the surround layer around it generates differential pressure changes such that it induces a higher density region of material on one side of an object and a lower density region on the opposite side. Some similar Magnus-like effect seems to act on the proton. An un-kicked ball cannot experience the Magnus effect, and this seems to be the case for stationary protons as well. Once a proton is moving the proton's magnetic moment would seem to provide the spin to induce the proton itself to be guided towards the lower pressure region that is formed. Which would explain the observed motion of positive particles through a magnetic field, as opposed to electrons passing through a magnetic field. With the electron, as always, seeking the denser path that it finds to be a better conductor.

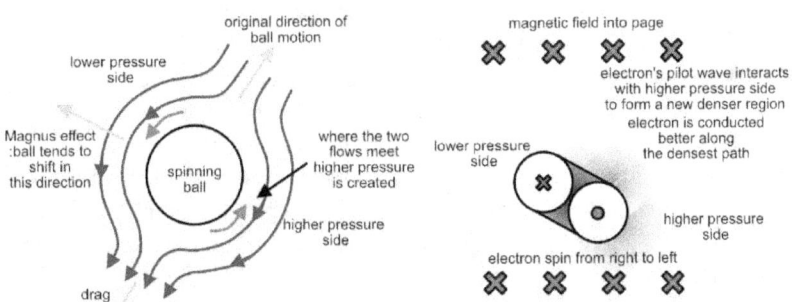

Figure 63- Magnus effect and its electron equivalent

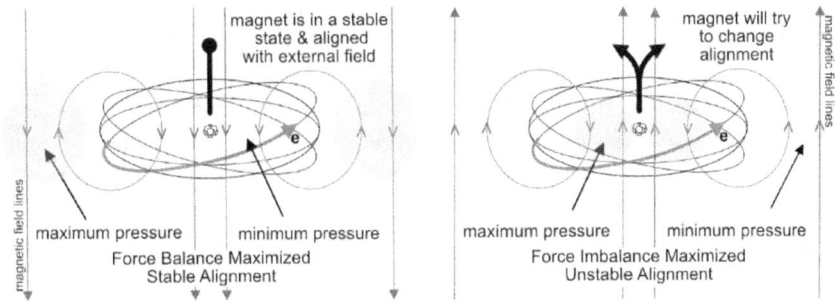

Figure 64-atomic magnet balanced and unbalanced systems

In a solenoid, the current or flow of electrons around a central core permits them to combine their rotational energies & focus a flow of FOS through and about it. Unlike the inside of the core, where the FOS flow is very strong, outside the core the fabric that is in motion is not focused and is, in fact, being dispersed by the action of the electrons and thus is comparatively weak, but readily detectable with the aid of a compass. The north pointer on a compass points in the direction of the FOS flow.

Referring back to our figures showing conventional current and electron current flows and their magnetic fields, we can deduce that when an electron is coming towards an observer, an electron rotates clockwise while of course inversely moving away it is rotating counter clockwise. [Neutrino spin supports this and was considered a violation of the principle of parity.(?)] But is it triggering angular momentum in the actual fabric of space itself, or do we simply see the sidereal translation of the motion of the fabric into circular motion by the electrons due to the nature in which they react and move with/cause the magnetic field. Back EMF seems to support the idea of angular momentum. And the re-normalization of the pressure differentials that exist outside the magnetic core, part of the mechanism of flow, would seem to support at least momentum. Consider plasma in a magnetic confinement bottle, or the Van Allen Radiation belts around the Earth. If it were just a matter of the conservation of the momentum of the fabric then we would not expect some of the behavior that we see in these plasmas for these situations. The plasma, charged particles, can bounce between the ends of the bottle and back and forth, and similarly we something akin to this inside the radiation belts formed around the Earth due to its magnetic field. For the radiation belts of the planets, is the aether/gravitational gradient's positive effect acting as opposing force/barrier to positive particles?

The resulting magnetic vortex appears to circulate the FOS around and through itself, and does not just suck up some of the fabric from in front of itself and shoot it out from its north or south poles. The electrons are moving a very small amount material, relatively speaking, from outside the core to the inside, and a pressure differential is also likely activated that induces motion from the front to the back of the magnet. Consequently, the fabric is set into motion about and through the host atom whose electrons are triggering the formation of a magnetic field. The apparent, or effective, circular motion taken about and through the magnet is brought about, of course, by the rotational energy (spin) of the electrons.

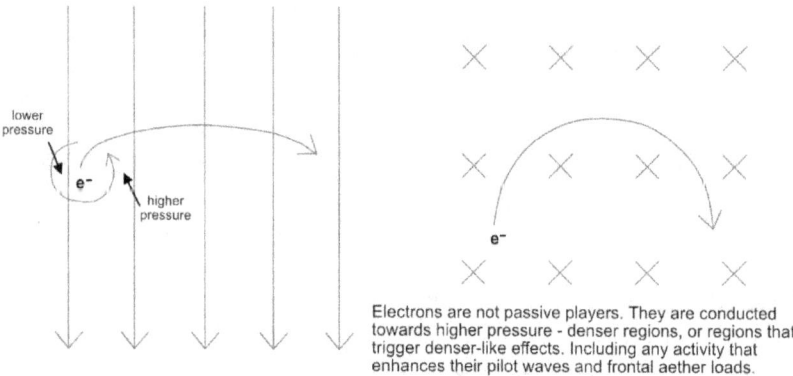

Electrons are not passive players. They are conducted towards higher pressure - denser regions, or regions that trigger denser-like effects. Including any activity that enhances their pilot waves and frontal aether loads.

Figure 65- electrons are not passive

Electrons are not passive players as we can see from formation of magnetic fields around electromagnets, and they are conducted towards denser aether/FOS gradients. The way in which electrons interact with magnetic fields supports the basis of electrons conducting themselves against the flows of magnetic fields, and any activity that enhances or increases their preceding pilot waves and frontal aether loads conducts the electrons in these directions. In total opposition to the way positive particles behave, which seem to act as if they experience the Magnus effect.

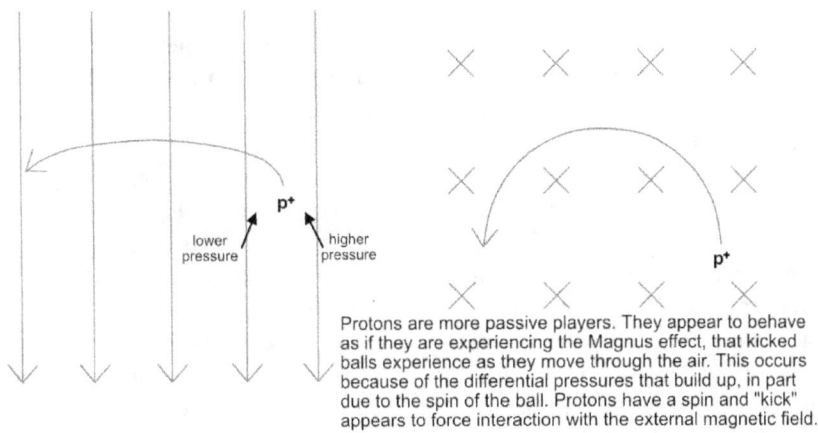

Figure 66- Positive ions are more passive

Positive ions in general move counter-clockwise and downwards when entering a magnetic field where the field is oriented downwards. They are more passive players compared to electrons and similar to a ball they cannot react until kicked or thrown. Is this a variation on the Magnus Force or Bernoulli's Principle? Their atomic electric field orientations are likely affected by the apparatus that sets them into motion and propels them into these magnetic fields. There is usually an electric field used to drive them into the magnetic fields. The spin and movement of protons is more complex in more dynamic situations like the Earth's magnetic field, where it also has gravitational/positive effects influencing them.

Many people at this point might make the mistake, that I did at first, in assuming that it is obvious that the FOS vortex and its inferred liquid characteristics are responsible for the How's & Whys of magnetic attraction and repulsion. That is if one tried to see the vortices as if they were simply composed of a liquid. I realized the mistake when I attempted to fit the characteristics of a fabric flow to the apparent effects of magnetic fields. The problem was that - "Why then did non-magnetizable materials, such as plastic, not move at all in the presence of a magnetic field?". If the flow was so strong, then one would think that any matter placed in the vicinity of the vortex would be set into motion. Only some of the aether has been induced into motion and inducing some pressure differential relative to its surroundings. (Consider an experiment of placing non-magnetic materials in the vicinity of a magnet where the effect of gravity is negligible.) Unless of course, you consider that the flow might appear extremely weak to all but the least secured electrons. [Consider the idea that a weakly held electron sees the motion of an electrical gradient moving towards it as being equivalent as its motion towards a stationary electric field.] For example - a changing magnetic field can induce electrons into motion in metals, where they are weakly held (metals are in essence supposed to be like a "cloud" of positive ions afloat in a sea of electrons), thus resulting in a flow of electrons in the form of an electric current.

The generated vortices are not directly responsible for the physical manifestations in the interactions between magnetic & magnetizable materials. It is likely in part that they play a role, but it is always predominantly the de-pressurizing wakes of the electrons that force the above fore-mentioned materials to interact the way they do. In magnetizable materials, the flow from a magnet will initiate those electrons capable of alignment & motion into forming magnetic cells of their own. The mechanics of angular momentum alone will not set any atoms into motion. It is the formation of a negative pressure "halo" near the parent nucleus by a pair (group) of electrons that induce the entire atom to move. When this "halo" is formed, it is often induced to move just off center, parallel, to the plane occupied by the nucleus (or at least its center of mass) by the lateral motion set into its parent electrons. [What role does the magnetic moments of the protons and neutrons is yet to be determined with respect to this new model.]

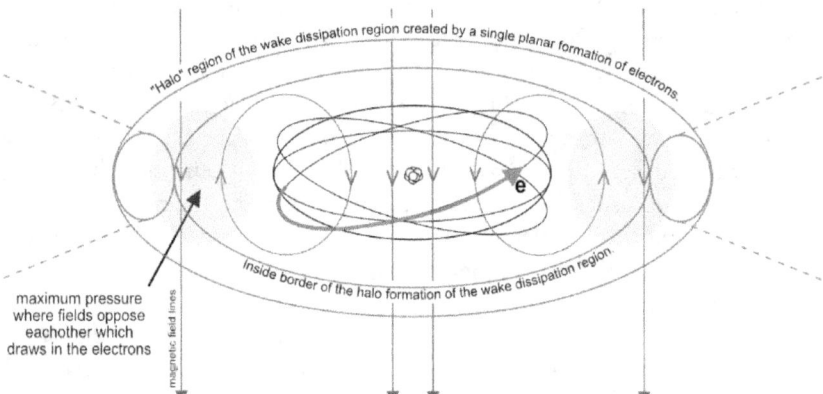

An electron involved in a permanent magnet takes up a roughly single planar orbit, and is electrically constrained about this orbit around the nucleus. External magnetic fields that are not of the same orientation will trigger a shift in the alignment of the electron as it adapts, and shift its generated "halo" region. This will alter the position of its host nucleus which reacts to the shift in the position, and pressure of the wake dissipation region.

Figure 67- halo region created by single planar orientation

This action having been set into motion by the energy of the fabric, from the vortex the electrons created, [in trying to carry out the conservation of angular momentum with the neighboring vortex that] has induced all of this. Thus, the positive (high-pressure gradient) nucleus is drawn by the wake region, which has developed slightly off to the one side of the nucleus, forming a negative (low-pressure region) halo. Or to put it another way, which is more technically accurate, the nucleus is forced towards the low fabric pressure region of the "halo" by the pressure place upon it by its own FOS gradient pressure from the side opposite the "halo". We also have in the separate magnetic bodies the interaction of the pressure gradients generated by the action of the electrons with each other, and forcing the electrons to react with them in some instances. For electro-magnets we also have the lower pressure regions possibly attracting the nuclear gradients on their own.

Just as electrons sometimes force nuclei towards "gravitational wells" (FOS density gradients) we have now discovered that the electrons forming magnetic fields (vortices) can also draw matter towards other magnetic fields via the same principle of the negative (lower pressure) wake region produced by wakes of electrons.

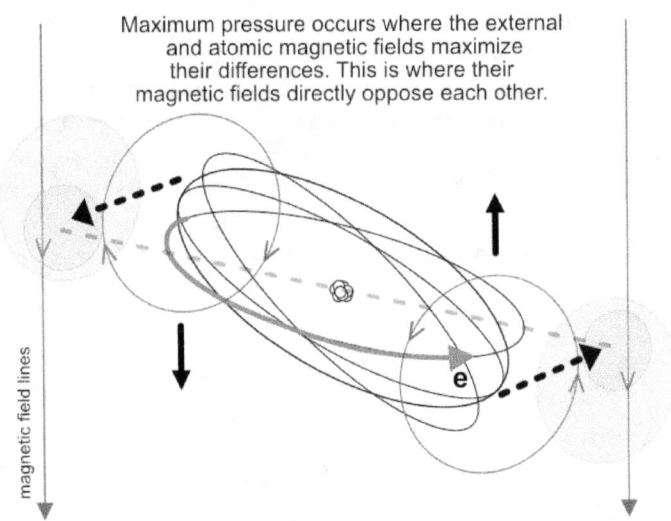

An electron that is part of a permanent magnet system will adjust its single planar orientation about its orbit, around the nucleus, to align itself perpendicularly, with any external magnetic field that is strong enough. This is triggered by the changing locations of the maximum pressure regions that form where the magnetic fields of the two meet and maximally oppose each other.

Figure 68- atomic magnet changing orientation with external magnetic field

The material ahead of the electron, both in terms of its frontal-load and pilot wave, play a critical role in the interaction of magnetism in two parallel current carrying wires. As well as in electrical arcs and lightning bolts.

[More on this in the future second edition of the book.]

Magnetism in Two Current Carrying Wires

When two current-carrying wires are placed parallel to each other, their magnetic fields interact with one another in one of two ways depending on the direction of the current flow relative to one another. Note that with parallel wires, we have field lines that are circular around the wires while with permanent magnets, we have vortex-like formations of field lines. From electrical arcs, and lightning bolts we know that the presence of a conducting wire is not necessary for electrons to flow together. Therefore, when it comes to a wire it just happens to help with low energy electron flow, low voltage events, to provide a supporting structure for them to move together and thus share a common path. Let us first look at the next figure where we can see that when the electrons are flowing in the same direction, their magnetic fields between each other oppose one another, and thus this triggers the electron-opposite Magnus effect, which results in the wires being drawn together. We must remember that when it comes to the nuclei, only the pressure balance between its opposing sides matters in terms of a driving force. These "free" electrons, guided by the wire, are drawn to the side where the more conductive fabric brings them, and they, in turn, draw with them the nearest nuclei in their vicinity by the shift in the position of the lower pressure dissipation wake region. To which of course, the positive nuclei are pushed towards by the differential FOS gradient pressure surrounding them.

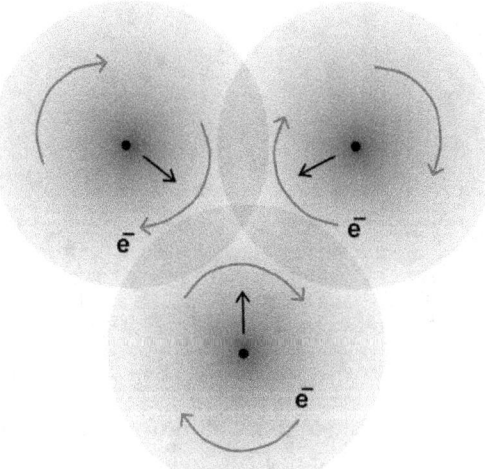

Figure 69-electrons traveling in the same direction attract each other by increased denser regions being created by the opposing magnetic fields and the Magnus effect

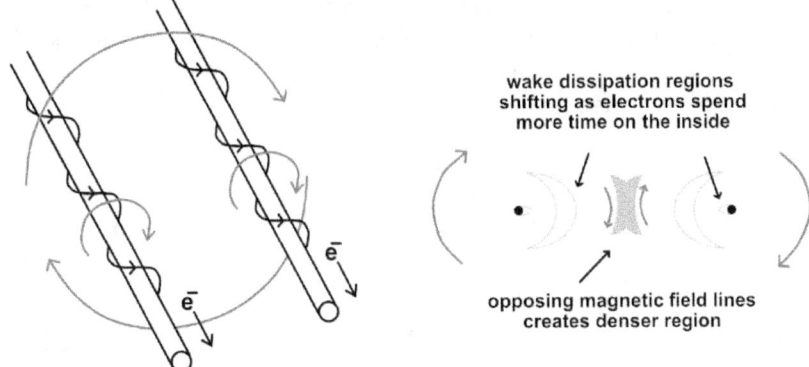

Figure 70- electron current flow in two wires in the same direction

Here we are looking at two wires with the electron current in the same direction. The wires are drawn towards the common center as the electrons shift their orbitals towards the denser region created by the opposing magnetic field flows, and by doing so shift their dissipation wakes as well.

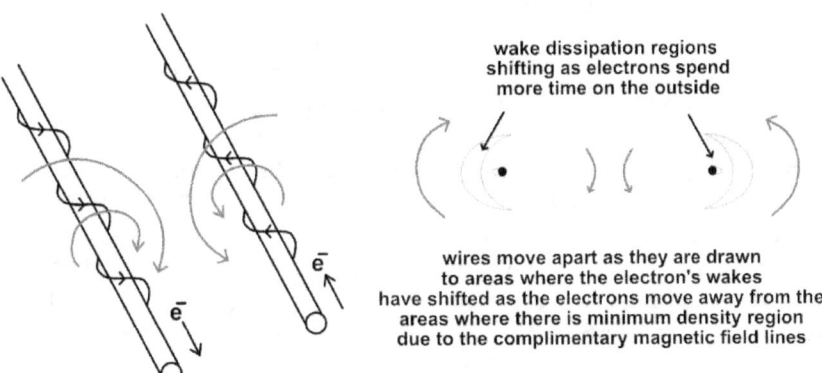

Figure 71- electron current flow in two wires in opposite directions

Here we show two wires with the electron currents traveling in different directions. The nuclei are drawn out of the area between the wires, and away from each other, as the electrons spend more time outside of here as well, and thus shift their wakes to spending more time here.

The Right-Hand Motor Rule

The right-hand motor rule is a physical representation by use of the right hand to indicate in which direction a wire will be "pushed" when it is carrying a current and in a magnetic field. Such as one that might be found within a motor - like that represented in our diagram. The thumb, index & middle fingers are extended in such a way that they are all perpendicular to one another. If the index finger points in the direction of the field (FOS flow) between the poles of a couple of magnets, and the middle finger points in the direction of the electron current flow in the wire - then the thumb points in the direction the wire is drawn towards.

right-hand motor rule

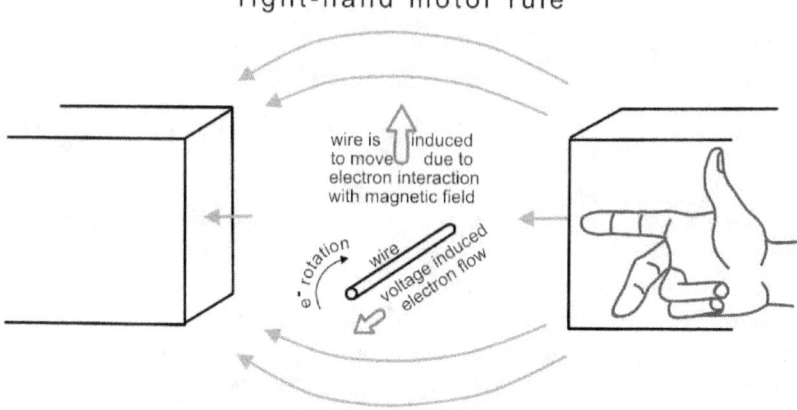

Figure 72- right-hand motor rule

What is happening is similar to what happened with two parallel wires with opposing currents flows. The electrons are forced into spending more time on one side of the wire because the opposing magnetic fields around them shift the electrons into a more eccentric "orbit". On their journey around the wire, the electrons travel more due to this greater path length. The result is that since the electrons spend most of their time to one side of the wire, the FOS pressure is lower here than that of the opposite side of the wire. Thus, the FOS gradient pressure on the nuclei of the wire pushes it out of the magnetic field in the direction indicated by the thumb.

[More on this in future second edition of the book.]

The Left-Hand Generator Rule

The left-hand generator rule is a physical representation by use of the left hand to indicate in which direction a current will be induced to flow when a wire is forced to move within a magnetic field. Such as one that might be found within a generator - like the one represented in our diagram.

The Death of the Dark Energy Idea

left-hand generator rule

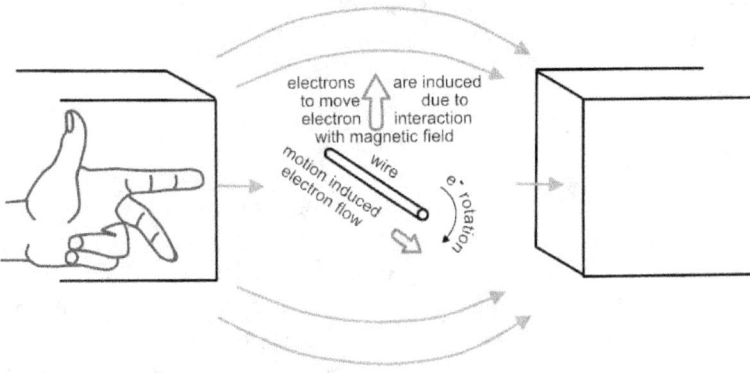

Figure 73- left-hand generator rule

The thumb, index & middle fingers are extended in such a way that they are all perpendicular to one another. If the index finger points in the direction of the field (FOS flow) between the poles of a couple of magnets, and the thumb points in the direction the wire is mechanically forced - then the middle finger points in the direction of the current that will be induced. Which is, in this case, the direction that the electrons will be propelled along the wire towards the reader. They are ejected out of the magnetic field (FOS flow) of the magnets, along the wire, for the simple reason that when an electron is loosely held, by some nuclei, and this metal system is set into motion, then these electrons are aligned & set along this path by these forces. Now that they are aligned, and forced to remain so as the wire is in motion, then electrons have no choice but to be conducted in the direction & only path left to them. Thus, they move along the wire & out of the magnetic FOS flow's influence.

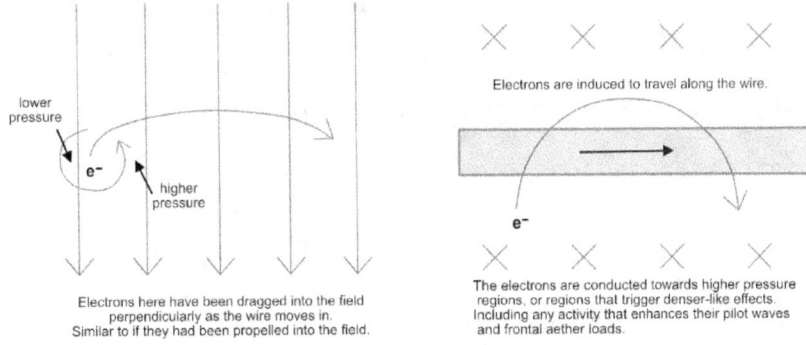

Figure 74-electrons induced to move along a wire

When the wire is not in motion then the momentum of the FOS flow of the magnet does not possess enough energy to affect the FOS of the frontal material of the electrons that are weakly held. Thus, they cease their alignment and are free to move about their respective nuclei as the surrounding FOS gradients & electron populations permit.

[More on this in future second edition of the book.]

Motion in the Earth's poles

When completely free, no nuclei governing their motions, electrons in a magnetic field will move against the flow of the fabric - along "the lines of force". Positive ions tend to move with the flow (& the drift). The electron's own fabric flow's forced assimilation to the angular momentum of the greater vortex results in the electron being oriented so that its only direction of propagation is against this drift component[?]. This drift component being initiated not by the rotation of an electron, but by the momentum gained by the fabric from being carried along with any electron in the intake of one of these smoke-ring (torus) shaped waves/particles.

[More on this in future second edition of the book.]

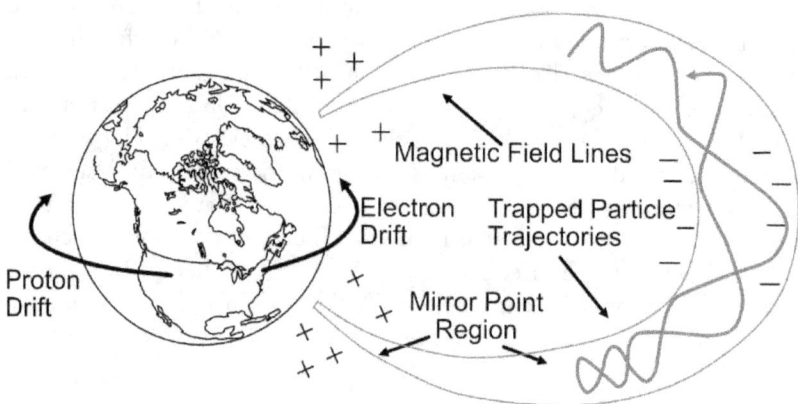

Figure 75- Proton drift vs electron drift with mirror points

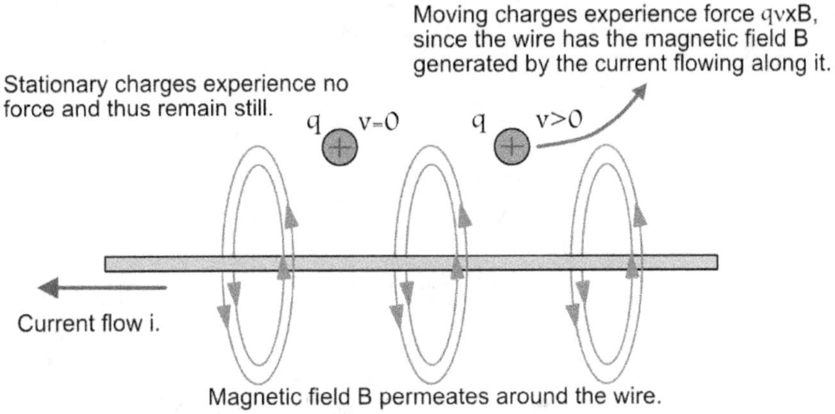

The Death of the Dark Energy Idea

Moving charges experience force qvxB, since the wire has the magnetic field B generated by the current flowing along it.

Stationary charges experience no force and thus remain still.

q v=0

q v>0

Current flow i.

Magnetic field B permeates around the wire.

Figure 76- Magnetic field B and charged particle interaction around a wire

Chapter Six

PHOTONS & ELECTRONS

But you don't have to be an ichthyologist to know when a fish stinks.
Unzicker, Alexander (1965- present (2023))

Photons – electromagnetic waves

The basis of the formation, growth and/or the production of all the chemical compounds, crystals and life around us has in one way or another has been initiated, or affected, by the interactions between electrons and photons. Without the warmth and the light of the sun, the electrons from the matter from which we evolved would never have had the energy, on the surface of the Earth, to be initiated into action to produce photosynthesis and allow us to have arisen from the oceans. [Disregarding chemosynthesis at this time.] Energy from photons was needed by electrons to have carried on the necessary chemical & physical reactions which sculpted the surface of the Earth and inevitably to produce us. Not only from photons forcing electrons into higher more sensitive orbital positions to allow chemical reactions, but also to trigger thermal agitation allowing water to evaporate and give us rain. And many other reactions.

The story of our understanding of photons begins with René Descartes (1596 March 31– 1650 February 11) and his proposal of a corpuscular theory of light. Later supported by Sir Isaac Newton (1642 December 25 – 1727 March 20). While the classical wave theory of light was promoted by Christiaan Huygens (1629 April 14 - 1695 July 8). With Thomas Young (1773 June 13– 1829 May 10) bringing back the apparent wave-nature of light with experiments using water waves (transverse waves), and the introduction of the single-slit and double-slit experiments. The failure in his reasoning that set the others on a path that they were unable to recognize - is that it's the water's displacement, of the medium conducting it, that allows surface waves to collapse in and out of the peaks and troughs they create to "turn" corners. Not a property of transverse waves themselves. This is seen in sound waves, which are longitudinal waves, as well as they travel down hallways and through doorways. After Young, this was then followed by critical experimental work by Heinrich Hertz (1857 February 22 - 1894 January 1) proving the existence of electromagnetic waves theorized by Maxwell, the nature of light was re-evaluated by Max Planck (1858 April 23 - 1947 October 4) and then finally took on its final form with help from Albert Einstein (1879 March 14 - 1955 April 18) with his work on the photoelectric effect. With Max Planck being the originator of quantization, and what eventually became Quantum Theory. Where quantum refers to the minimum quanta or quantity that could exist expressed in the equation $E = hf$. Where h became known as Planck's constant, also known as Planck's action quantum. The one player which applied wave theory to matter was Louis de Broglie (1892 August 15 - 1987 March 19), and his abandoned, until recently, the theory of pilot waves. Sid Deutsch introduced the idea of pilot waves to me shortly before his death. There is no transverse wave model that has the mechanics to trigger pilot waves, but it is a natural occurrence of compressional longitudinal waves. This sometimes minor, but critical compression-density wave is the material ahead of the photon that is…more on this in the future second edition of the book.

[Add image of pre-compression forming ahead of wave. Image being updated.]

The classical wave theory of light, based on transverse waves, could not explain and was, in fact, in direct conflict with the observational data around the photoelectric effect. It, in fact, predicted that all fires should start out as what we see as the normal light of fires but then should quickly turn into a fire dominated by the higher frequencies of light in the ultraviolet end of the spectrum. At the time x-rays and gamma-rays were not yet known, and had they been known then the theorists would have predicted them arising out of a normal fire.

Not only did this not happen, obviously, but it also did not predict the existence of a threshold frequency below which the photoelectric effect could not work, and above which electrons where ejected. On top of that classical theory predicted that the intensity, or brightness, of the light source, would matter, but it did not. Only by reaching the threshold frequency, density, could an electron be ejected and furthermore it did not predict the increasing kinetic energy gained by exposing electrons to higher and higher frequency light.

Our modern understanding of light begins with what was perceived as a mystery regarding fires and the spectrum of black body radiation. At high temperatures, most of the radiation should be in the ultraviolet region, and beyond. The idea was that all frequencies were equally probable since there are more high frequencies than low ones. It should be more likely that the frequency of light should be emitted from the high end of the spectrum. But that is not what is seen. The ultraviolet end of the spectrum instead drops off to zero. Some scientists called this the UV Catastrophe. Because predictions based on classical wave theory, with transverse wave mechanics, were so different from what was observed. Scientific theory and experimental observation were at odds. Obviously, the classical theory, the transverse wave model, was wrong, but why?

Max Planck's Quantum Hypothesis

Planck started by investigating the observed graphs of black body radiation sources and worked backwards until he found an equation. $E = nhf, n = 1,2,...$ But he did not fully understand why it worked. Planck hypothesized that the atomic oscillators that produced the observed radiation had two special characteristics. The oscillators had energies related by the equation $E = hnf$ [$n = 1,2,...$] Where f was the frequency of the oscillator and h was some constant. Eventually, h became known as Planck's constant. So, the atomic oscillator was somehow restricted to having energy related by the discrete integers 1,2,.. and so on. But an oscillator could not have fractional energies of n. If an atomic oscillator lost energy, then light would be emitted based on losing energy from one energy level to the next. Planck's idea explains why it is much more likely why the atomic oscillators would emit low frequency red light before emitting higher frequencies like ultraviolet light. Higher frequency UV light requires more energy, but before the oscillator has collected enough energy to emit UV light it is more likely the energy will already have been released by the lower frequencies infra-red, red, orange,.. light.

The quantum is the lowest amount of energy that an atomic oscillator can have. Which is Planck's constant times a frequency where $n = 1$. All other energies allowed are some multiple of this fundamental value. Is this due to the granularity of the FOS/aether?

The classical theory of physics said that a vibrating object even an atomic oscillator could vibrate at any frequency. Planck's theory said that for black body radiation that classical theory did not apply and that only certain energies were allowed. Planck at first thought he had stumbled upon some mathematical trick and for several years he tried to find a classical basis for his quantum hypothesis. For his discovery, he was awarded the Nobel Prize for Physics in 1918.

The photoelectric effect reinforced the theory of Planck's quantum hypothesis. This effect was observed when photons strike the surface of a metal and trigger the ejection or release electrons. In the experimental apparatus, the ejection of these electrons would then allow previously charged thin metal strips [called leaves] or bars to come back together again. Classical theory also predicted that their should be a time delay as the frequencies had to add up to trigger the ejections, but no time delay is observed. And although a UV light would trigger ejection right away and increased with the intensity of the light, while red light, for the given experiment and metal, regardless of the intensity of the light or the time shining on the metal plate did nothing. This is because the light has to have some minimum frequency above the minimum value to eject an electron. This minimum value is called the threshold frequency for the particular metal, or work function.

One of the key experiments was done by setting up an electrically-isolated polished-zinc metallic mass with metallic leaves at the other end of it. Now give the zinc body a negative charge, and the leaves will spread apart due to their negative charges which causes them to separate. The electrons will be discharged, and the leaves fall back together, in the presence of UV light. Any light of a lower frequency/density and the leaves will stay apart. The zinc mass will stay charged. Heinrich Rudolf Hertz was the first to observe this phenomenon, in 1887, while he was verifying the classical theory of electromagnetic waves. Metal surfaces lost their negative charge when exposed to ultraviolet light. Using a red-light source, and increasing the intensity of a red light had no effect, because it was below the metals' threshold frequency. The photoelectric effect depends on the frequency and not on intensity/quantity. Classical wave, transverse, theory predicted that intensity would play a roll, and it did not. The higher the frequency correlates to a higher aether-photon density.

The classical wave theory of light had been brought into question by black body experiments. If the classical theory had been right, an intense red light for longer periods of time should have triggered electron ejection, but it never did. Only photons at or greater than the threshold frequency can trigger electron ejection, and there is no time delay. It happens right away if the frequency is high enough or not at all if the frequency is below the threshold value. On top of that, as we increase the frequency past the threshold values, the electrons are ejected with more and more kinetic energy. The intensity of these frequencies does not increase the kinetic energy it only increases the number of ejected electrons. Again, conflicting with the classical theory of light. The transverse wave model.

It was then Einstein who re-evaluated the data and observations to propose his photon-of-light theory. For which he won the Nobel prize in 1921 sixteen years later after the publication of his paper in 1905.

By utilizing Michael Faraday's study of electric and magnetic fields James Clerk Maxwell developed his equations, for electromagnetism, which have allowed us to understand that a photon is an electromagnetic wave. Having both an electrical and magnetic component. He was able to predict the existence of radio waves and that they would travel at the speed of light.

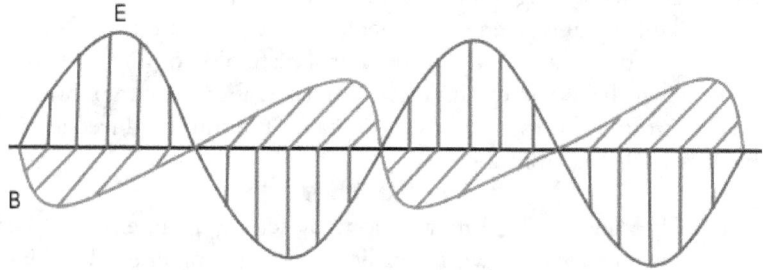

Figure 77- typical representation of an electromagnetic wave

The figure above shows the standard representation of an electromagnetic wave as a transverse wave. No restoring force is ever mentioned. Yet all other transverse waves require a restoring force to continue to exist, and propagate through the medium conducting them. Maxwell realized that a changing electric field creates a changing magnetic field which creates a changing electric field… thus keeping each other going. But what are these at a structural level without putting off the explanation to some other particle or hidden force that we are only able to imply by the effects we can observe?

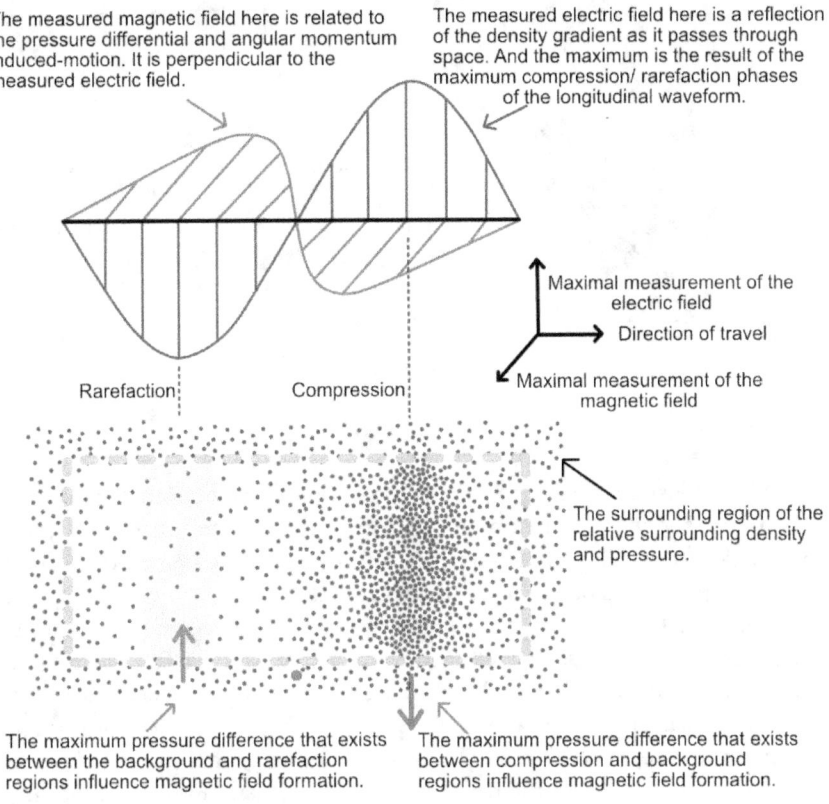

Figure 78- Transverse wave form profile from a longitudinal wave body

The image above shows exaggerated varying density, with compression and rarefaction, for easier portrayal and clarity. The following older images are being used to explain the longitudinal density-gradient wave nature of a photon.

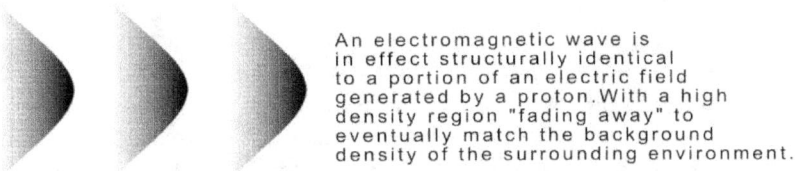

Figure 79- a compression wave is equivalent to a gradient body

In the figure above we have a FOS density gradient representation of a series electromagnetic waves. A compression wave is by definition a gradient, and thus an electric field. The movement of which triggers a magnetic field due to the effect the photon has as it passes through space.

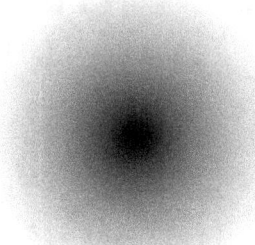
An electric field around an atom is in effect an aether density gradient of a high density region, being generated by the proton, that then weakens until it blends in with the surrounding aether. A photon that is of similar density can penetrate to a matching density. This is why the photoelectric effect exists, and e⁻ activity within gradients creates layers to give us quantum mechanics.

Figure 80- Atomic gradient is an electric field

First, consider the standard representation of an electromagnetic wave as in the figure showing the mapping of a longitudinal wave form into a transverse wave-like form. The wave portrayed is not a standing wave but is due to the measurement of its properties as the waves pass through a region of space. Now consider the figure showing multiple waves passing through space, and think of it as a continuous passage of electric gradients. We need to consider the angular momentum of light that Richard A. Beth showed us is detected in a doubly refracting slab showing that photons possess some component of rotation. The reaction torque, experienced by such a slab showing that photons have angular momentum, is a clue to the true nature of magnetism. Previously we suggested that the rotation of electrons must be what induces magnetism by inducing motion of the aether it passes through, and now it's time to consider the same thing for photons. A photon's gradient and motion through the fabric must also induce some component of motion simply by the nature of compression and rarefaction of the aether. De Broglie's pilot waves are present, and could be the source of what are often referred to as virtual photons.

A photon can be produced when an accelerating electron ceases to accelerate, in one direction, releasing in the process any excess fabric that it has compressed ahead of itself during the acceleration, and which its change in direction and velocity free the compressed material. Which is how radio waves are initiated as well as synchrotronic radiation. Thus releasing/emitting from the intake of this the electron, a volume of compressed fabric which in turn is shed and transfers its motion-gained energy to the fabric it had been moving towards. Causing the compression & rarefaction of the fabric which is, by definition, exactly what a longitudinal wave is. (This kind of production of photons could be additional evidence of the formation of longitudinal photons.) Thus, a photon is born.

Is there a limitation on the density of a point in space that can be generated by the passage of photons through it, from whatever number of directions they can come from, since a photon is a momentary high-density point generated within the surrounding region's aether relative to the surrounding normal density? In short, is there a limit on how dense a point can be generated? Due to a limit on how much the fabric can be compressed in a given region. This would produce an upper limit and restrict certain types of actions or reactions. Any region that became too dense would simply disperse in a shower of gamma-rays, and other forms of electromagnetic radiation.

Erratic motion can also overcome the elasticity of the extra fabric ahead of an electron such that it will be unable to also change course with the electron due to its accelerating in a new direction. Allowing the momentum of the extra fabric to tear itself free from the electron, and become a photon. The Electron Wiggler Laser is a device which alternates between accelerating electrons and forcing them to undergo rapid changes in their direction of travel and lose energy through the formation of photons. And thus because of this alternating motion, the term wiggler is used. The greater the acceleration (the greater the compression) the shorter the wavelength & thus the more energetic the photon that is produced. While inversely, of course, the more limited the acceleration (the lower the compression) the longer the wavelength & thus the less energetic of a photon is produced. Of course, how rapidly the electron slows down is going to affect the expansion of the fabric it contains within its intake or cone-like depression. This behavior once again supports an aether.

More often the production of a photon takes place as an electron expends potential energy during the final stages of maneuvering from one electron orbital (such as 'M') to a lower or closer orbital position (such as 'L'). Or also in a chemical reaction where an electron from one atom moves to a more conductive region about another atom. In the chemical reactions that we are more familiar with the photons normally produced are infrared - to most of us this is simply heat. Another common event is the collision or near-collision between electrons. This forced erratic maneuver jars from the front of the electron any additional FOS that has accumulated during recent accelerations. (Given that an atom has a specific gradient body that it maintains, do electron orbital transitions disrupt the balance, and instead of just adding additional more-dense material the atom's gradient cannot support it and thus it must be ejected/diverted out of the atom's gradient?)

Photons affect electron conduction quite simply by presenting themselves as regions of denser than normal fabric. And thus, just as FOS gradients, they present themselves as more conductive regions within space. The electrons have no choice but to react to and absorb this energy and be diverted out of the orbital shell it inhabited, just as it follows any conductive FOS gradient presented to it. In the process accumulating the fabric of the photon within its vortex intake. So, no energy is lost it has merely changed form. Perhaps even giving the electron a chance to travel to another atom, and possibly initiating an electro-chemical reaction, which would result in bringing the two atoms together.

The larger the photon (greater the wavelength) the less dense the wave (region of space), and thus the lower the potential acceleration rate away from where the electron had been. And vice versa. We have the photoelectric effect.

The Photoelectric Effect

The photoelectric effect is due to the tiered density layers formed around nuclei, and a photon's ability to penetrate, or not, to some density layer. Which is why classical electromagnetic theory, could not account for the behavior of the experimental data. Since any photon less dense could not penetrate and thus not trigger the ejection/diversion of electrons. Albert Einstein explained the experimental data of the photoelectric effect as the result of light energy being carried in discrete quantized packets [1905]. And thus, quantum mechanics was reborn. [Or should it be said that Planck's QM was supported and extended?] The photons had to be of sufficient energy-density, and anything less could not penetrate and eject an electron.

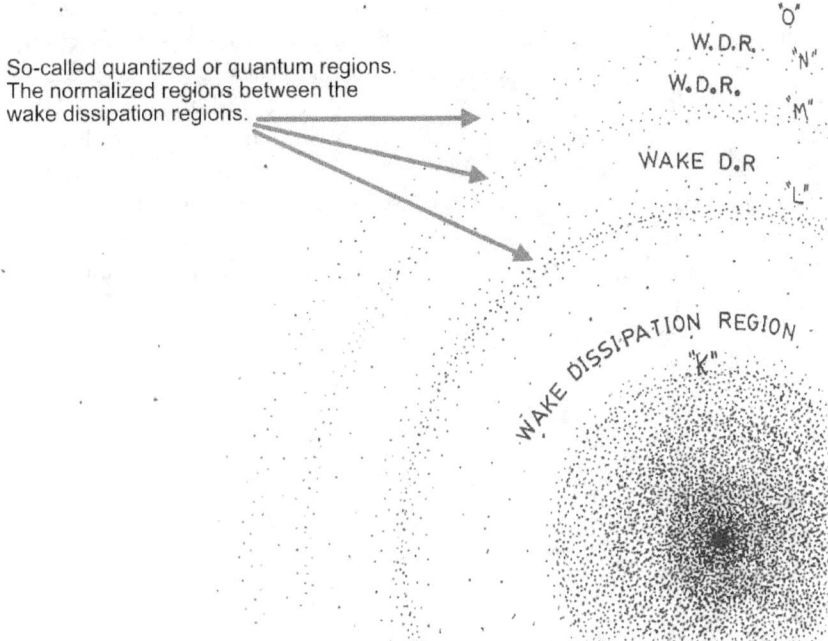

Figure 81- wake dissipation regions around an atom. Exaggerated scaling.

The wake dissipation regions triggered by the electrons orbiting atoms give them their tiered density layers: the denser a photon, the greater its penetration capability into an atom. The layering may be more linear in nature than implied by the figure. All that really matters is that the regions separated by an electron above another simply lack enough of a gradient-change to draw in and keep another electron. But to an electron at the outermost shell, it would more likely be kept in orbit due to a true drop in density relative to the density around it, and the gradient core that is more attractive to it.

The tiered density layers are due to the wakes, made by the electrons, and the attempted reformation of the density gradient after the passage of electrons through it, and the dissipation of the wakes. A sufficiently dense photon entering into the FOS gradient around an atom presents itself as just as an effective conductor to the electron as the gradient it occupies and thus allows the electron to be routed away. A photon not dense enough cannot penetrate any further because the material closer to the nucleus is denser and thus it's reflected back out. Each level (orbital or shell), after the 'K' shell, within an atom is formed where the fabric has just returned to a sufficient potential for conduction. Just beyond the 'K' shell's electrons' wake dissipation region. Where their wakes have just diminished enough in energy so that this newly restored region is just of a great enough density that it is capable of conducting electrons indefinitely at this radius from the nucleus. This region then permits (due to volume & surface area) a group of electrons to take up residence in a hierarchy of positions (levels) based upon how each electron's conduction strength affects the fabric immediately around it via its wake. Which in turn lowers the fabric conduction strength above it and thus lowers the next electron's strength of "attraction". And so on. The energy of all the wakes eventually reaches a series of peak expansions in their decays, preventing any electrons from inhabiting the said region and simultaneously defining the border of the next potential shell. What each potential orbital, the region inhabited by a single electron, represents is a layer of fabric of a unique (for that atom) density.

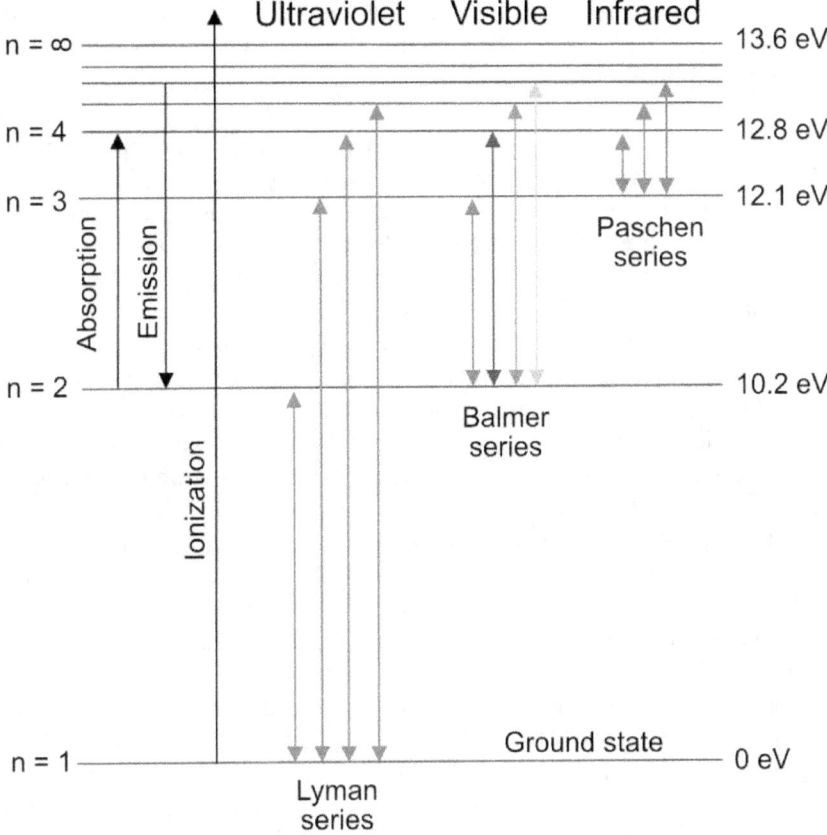

Figure 82- Lyman, Balmer and Paschen series for Hydrogen

Consider the hydrogen atom. The ionization energy of the hydrogen atom is the minimum energy required to trigger the ejection of its only electron, per mole, in its gaseous state under standard conditions. That is at a temperature of 298 Kelvin and a pressure of 101 kPa. This ejection energy corresponds to the difference in energy between the levels n=1 and n=∞.

The minimum wavelength responsible for the ejection of an electron, out of what might have been a stable orbit, is due to the fact that this photon must be at least the same density of the very fabric that the electron inhabits. And thus, appearing to the electron as good, if not better, a path of conductance as any previously traversed region of the shell it was within. In other words, if it takes an ultraviolet photon to cject an electron out of an orbital then the density of the fabric of that orbital is at least equivalent to the density of the ultraviolet photon. Thus, the correspondence between the emissions[16] and absorption frequencies of the atom. For the hydrogen atom, this corresponds to approximately a 94 nm photon from the Lyman series of the spectrum with an energy equivalence of at least 13.6 eV.

The photoelectric effect takes place because if the photon is not of a great enough density, then it will be merely reflected back, by the more dense fabric around the atom. Blocked before ever reaching a layer of FOS that most electrons inhabit. Photons whose wavelength are too long (too low a density) are incapable of ejecting (leading) an electron out of an atom. But may trigger orbital distortions in the more weakly held outer electrons which is seen as thermal agitation, and thus such atoms see an increase in their temperature due to emissions in the Paschen series part of the spectrum, or perhaps the emission of a visible photon from the Balmer part series of the spectrum.

The photon-electron interaction, coupled with the production of photons through electron collisions play key roles in the inhibition of ultraviolet photons from regular fires. This so-called absence of ultraviolet photons phenomenon being considered one of the two so-called Ultraviolet Catastrophes, also called the Rayleigh–Jeans catastrophe. What they were catastrophic to were the then so-called modern-day theories on structural atomic physics, starting with Christiaan Huygens support of the wave theory of light and culminating in Einstein's photoelectric effect paper in 1905. All of it starting with the Classical Theory of Light. According to one of these theories, if correct, ultraviolet photons should be emitted by any fire. And in fact, due to the harmonics of transverse-wave photons they should be followed by the emission of x-rays & gamma-rays from the same fire. At the time of the development of the classical theory x-rays and gamma-rays were unknown to science. The FOS theory series is based upon photons consisting of longitudinal (compression & rarefaction) density waves, and not transverse waves.

The reason that these greater or more energetic waves should be emitted is that transverse waves can unite together to increase the amplitude, and thus the frequency, of the resulting waves that are formed by the uniting of wave peaks and troughs. [Need to add Max Planck information here.]

The reason that ultraviolet photons are not produced (according to the FOS theory series) appears to be that the energy of the electrons, in terms of acceleration, is too low and/or the orbitals of the nuclei in which U.V. photons could have been released are filled.

[Add more on the UV Catastrophe and accepted explanation.]

[Need to add images for clarity.]

Only when a point in the overall energy level (temperature) of a body of matter is reached are electrons capable of generating U.V. photons being

---

[16]actually I have not said anything about the emission of photons from an orbital positions yet.

released. [This could be wrong:-The electrons are accelerated by the "consumption" of the infrared energy along their paths, and can release that energy in collisions with other electrons or the attainment of a low orbit about some nucleus. :-end of possible error]

If the electron density is too high (low temperature), then collisions and the release of the extra frontal FOS bodies occur before the electrons have been accelerated enough to release U.V. photons. In U.V. photon production via orbital attainment & release it is important that:- a sufficient number of electrons have been cleared away from the accepting nuclei, and that the surrounding nuclei are also kept in the same condition. If these conditions are not met then other electrons from the surrounding environment & nuclei will move into the lower energy orbitals before more energetic orbital positions are exposed, and thus any available electrons will move into these unoccupied orbitals releasing less energetic (relatively less dense) photons. This usually occurring due to these electrons more limited accelerations. Thus, preventing the "freer" electrons from diving (accelerating through) into the potentially ultraviolet photon yielding regions of nuclei. The electrons which had the potential to yield U.V. photons would now then be forced to assume a lesser (higher) orbit. Thus, releasing a photon of an energy lower than that of a U.V. photon.

These are the physical (FOS) reasons that in regular fires, and in the heating of matter (furnace or light filament) that photons of around and above the ultraviolet range are not produced. In most cases, the bodies of matter in question would have vaporized, and dispersed, long before a U.V. fire was possible. As in most cases, it is necessary to remove from the nuclei of many of its electrons. The result being that on Earth ultraviolet light that we encounter, other than from the sun, is electrical and/or human-induced in origin. Too many electrons are usually nearby, and cannot pass up moving towards such a strong gradient/positive charge.

Two Slit Experiments

The two-slit experiments for photons, and electrons, have astounded people and under current modeling are deemed by some as a miraculous phenomenon. But as we will discuss using Louis de Broglie's, and Sid Deutch's, pilot waves they are just a common mechanical event related to photons passing close to the surface of some mass, and its gradient's influence. The two-slit experiments are just another instance of photons, and electrons, being affected by a FOS/aether gradient abrupt change.

It is astounding to learn that the "intellectual elite" have not considered the actual mechanism behind the spread of transverse waves on water after their passage through a gap in a harbour barrier. This spreading is not directly due to the transverse waves themselves, but is instead a result of gravity pulling down on the peaks of the waves outward, and on the surrounding sides into the troughs inward – to try and bring equilibrium back to the medium conducting them. A well-understood behavior by those who study water wave dynamics.

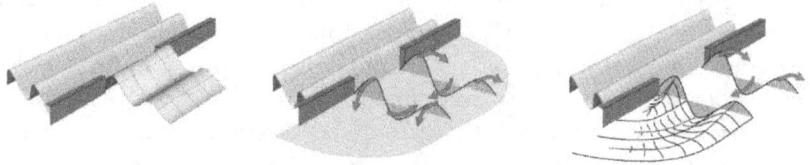

Figure 83- Reflection and Diffraction around Breakwaters

There is no component of transverse waves that allows them to spread out or twist on their own. Instead, it occurs because of the physical changes to the medium conducting them and not directly because of the transverse waves. Transverse waves can create the conditions in which gravity can pull down the sides into the troughs and pull down the peaks to the surrounding water level, but only then can the transverse waves follow the water's new path that it has taken. The transverse waves follow the water's surface as it deforms as it passes through barriers and/or moves around objects. If the side of a tank containing water collapsed, the water would spread out just the same in the absence of any transverse waves.

Additional evidence for a similar nature between photons and phonons can be seen in their diffractive behaviour after they have passed through a gap. With sound waves, it is the air that experiences pressure changes as the sound moves through it, and the pressure changes on the waves that move them around corners due to the drop in pressure as the waves "peek" around corners. In the following image, we see that higher frequency sounds bend less. Just as we also see in the adjacent accompanying image that higher frequency photons also bend less.

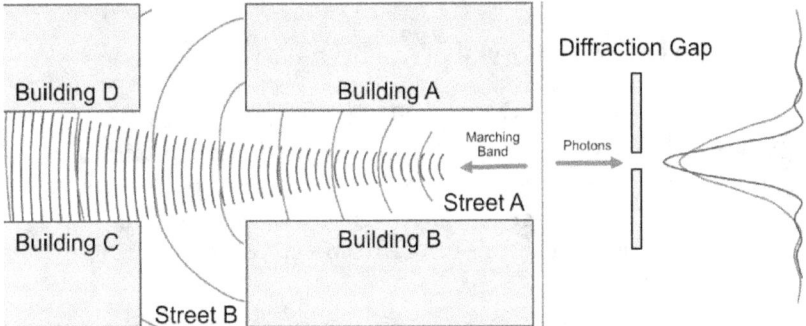

The longer wavelength sounds of musical instruments, of say a high school marching band, like the Tuba, diffract around corners more readily than the shorter wavelength, higher-pitched, sounds like those from another instrument say like the piccolo. Which presents itself by listeners coming down a perpendicular street hearing the lower frequency sounds more readily. Similarly, the longer wavelengths of red photons, diffract much more than the shorter wavelength blue photons where the gap separation is closer to the wavelength of the red photons.

Figure 84- Sound wave and light wave single slit diffraction

Apply this billowing out behaviour to the pilot waves traveling ahead of the photons, or electrons, to the two-slit experiments and we now have a mechanism for the so-called mysterious patterns observed. First pointed out by Louis de Broglie and then later by Sid Deutsch. Not mysterious at all and a simple in explanation as they follow the densest most conductive path ahead of them.

[Note: Look at showing the size comparison of red and blue photons in terms of wavelength, width, height and the volume relationship. Then look at the size of the profile and what size of pilot wave profile it would generate. The larger the more the billowing out after passing some barrier. And thus, the greater distortion of a photons potential path. Then perhaps apply this to electrons as well at least in terms of a descriptive image.]

[Perhaps the key to developing some of the mathematics lays in comparing the different wavelengths and their angle of change and relating this to the gravitation influence on photons. To develop a measure of the required density change or something akin to this.]

Phonons and photons show remarkably similar behaviour under similar circumstances. Including their speed being mediated by a conductive medium, whereas without a conductive medium as an explanation, Einstein concluded that we simply have to accept this fact and move on. In other words, since we can't figure this out let's just give up for the time being and hand wave it away. Nice.

[Note: Can simultaneously varying the material of the slits that the photons are passing through provide some kind of evidence of aether density influencing their paths?]

Now you see me, now you don't

The Observer Effect. The odd behavior of when you place a sensor to detect the passage of photons next to the slits and the interference patterns is reported to disappear or at least change to some degree. Which appears to be a sensor's proximity to the gaps triggering interference. Likely preventing the normal aether distortion similar to wave-gap diffraction.] [More on this in the future second edition of the book.]

The "Bucket" or "Nested Doll" Atomic Effect

Atoms act like buckets of FOS with denser buckets embedded even deeper. This bucket can only hold onto so much material of the same density, and thus any excess material is shed. And when this occurs in a nuclear fusion reaction this appears to us as the emission of photons and neutrinos. It is why in the fusion of light elements that large amounts of energy are released due to the relatively large change in density-volume profile. While in fission we see the establishment of more 'K' level, and other lower orbital electrons positions, that the larger nuclei started with and that simply are being broken up and rearranged.

If we consider Einstein's famous equation about mass and energy, then we can see that this mass to energy conversion is, in fact, the result of protons shielding one another from pulling in as much fabric as they individually could due to their standing-wave energy no longer radiating away evenly in all directions. These shadows reduce the compression of the fabric and thus reduce the apparent mass of the final nucleus in comparison to the starting masses of the individual protons and neutrons. If we can separate the nucleus back into the individual protons to give us hydrogen, then the mass is restored. The compressional radiant energy is once again at a maximum.

The proton count is constant, but the neutron count of each element in the periodic table has some wiggle room. These isotopes, whose neutron count varies for the same element, are often thought of just as a change in the neutron count with little effect on the chemistry of the elements. But that is not quite the real story when it comes to the simpler atoms like hydrogen, helium and others. Heavy hydrogen is not just heavier; its chemical behavior is actually different enough that water made with the most common form of heavy hydrogen, called deuterium, is actually toxic in low concentrations. Just 10% [*] of heavy water mixed with regular water is harmful to humans. And of course, tritium, with two neutrons, is radioactive with a half-life of about 12.32 years. Decaying into Helium-3 by beta (e^-) emission.

Property	H$_2$O	D$_2$O (heavy water)
Molecular weight,	18.015	20.028
Melting Point	0.00	3.81
Normal Boiling Point	100.00	101.42
Temperature of maximum density	3.98	11.23
Density at 25º Celsius	0.99701	1.1044
Viscosity at 50º C	0.8903	1.107
Refractive index	1.3330	1.3283

Table 14- Some physical properties of ordinary and heavy water (source lost)

Property	H$_2$	D$_2$
Triple point Co-existence of all phases	13.96	18.72
Normal boiling point	20.4	23.6
Critical temperature For liquefaction	33.24	38.35
Critical pressure	12.8 atm	16.4 atm
Heat of fusion	28.0	47.0
Heat of vaporization	216	293

Table 15- Some comparative properties of hydrogen versus deuterium (source lost)

If you think about the differences between regular hydrogen versus deuterium, this is evidence against the groups who argue that the physical constants cannot vary in the slightest or life would not be possible.

Property	Protium	Deuterium
Mass	1.007825032	2.014101778
Spin	½+	1+
Magnetic moment	+2.79285	+0.85744
1st Ionization (eV)	13.59844	13.602[?]
2nd Ionization (eV) (Electron affinity)	0.754195	0.754598

Table 16 - Other physical differences between hydrogen and deuterium (source lost)

The radiant area created by nuclear formation is the key to electron's carving out regular orbitals of ever-increasing size, but it is the electrons who do it. The nucleus just acts as a radiator of compressional energy that then has some specific rate of change of compression that either allows the electrons to take

up stable orbital positions, or triggers Electron Capture due to it being too strong. [I think we can add much more here on nuclei and maybe radioactivity.]
[Need to add images for clarity.]

Atoms and Atomic Structure

Atomic and molecular hydrogen. What role does atomic structure have in the boiling point of the elements, and what makes a gas a gas? Which is related to what triggers the boiling point of the elements.
[More on this in the future second edition of the book.]

Quantum Tunneling

Quantum Tunneling usually refers to the quantum mechanical phenomenon where electrons tunnel through an electric barrier between some groups of atoms, that by classical theory no electrons should be able to pass. This behavior is what most of our electronic devices in the world are based on, and these potential "tunnels" are enhanced when a small electrical field, or current, triggers additional distortions in the electron orbital positions and the density, or charge, gradient between atoms. The barriers are compromised or changed such that then they allow electrons to pass through between atoms more easily. The following image first shows that classical idea of how an electron would have to gain enough energy to pass through the electrical barriers between atoms, while in quantum mechanics somehow electrons are able to "tunnel" through these barriers. In our FOS model the explanation is obviously simple in nature. Due to the orbital precession of the electron motions the wake dissipation regions are also moving, and thus what is called a quantum barrier should be called the average quantum barrier with the average normally taken into account, but which directly neglects the minimum and maximum barrier values. With quantum mechanics using equations that can use the probabilities, and thus account for the "mystery" without taking into account why it occurs in the first place.
[More on this in the future second edition of the book.]

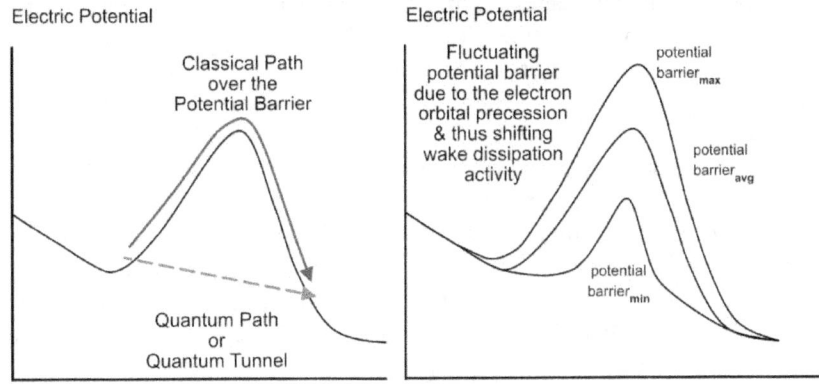

Figure 85-Quantum tunneling and wake dissipation effects

Superconductivity

What is superconductivity and what triggers it? To begin with by definition superconductivity occurs when the electrical resistance of a material drops to zero. And electrons can flow freely without interference. The other effect that occurs is that a piece of superconducting material placed above a permanent magnet will float above the magnet, or alternatively a superconductor can push away a magnet. This is said to occur because of the Meissner effect.

Superconductivity by its definition helps us answer the question of why it occurs in the first place. It occurs because connecting regions between atoms, and in some cases molecules, become effectively equal in nature to an atomic orbital region. Likely in most cases, particularly with higher temperature superconductors, as the temperature of a material drops to some critical point their electrons lock into a specific distribution pattern due to the crystalline structure that leaves some quantum-orbital-like regions, higher FOS density regions, between atoms and free of any electrons. Since they are in effect large regional quantum-orbital-like regions they are open to electrons to passing freely, zero resistance, through them. With these higher density regions extending between atoms then any applied electrical potential with a source of electrons produces a current with zero resistance.

Virtual Photons do share some common properties

Virtual photons are believed to be the carriers of the strong force, and some other interactions between particles. The aether gradient acts in some cases as the equivalent in the same way that a photon can represent itself as a better conductive path to an electron. But not for the strong force. The strong force works quite differently and in an opposite manner to the protons. It is the absence of material, lower pressure, that draws in the protons to the common center, while when a greater amount of material lays between a pair of protons, they are actually being pushed apart.

In magnetism the pilot waves may add to its power due to the induced motion, and sometimes magnetic fields are described as circulating as virtual photons. [Who said this?]

[More on this in the future second edition of the book.]

Interference & The Anthropic Principle

Interference patterns among photons can produce some quite stunning patterns of light when they involve a variety of different wavelengths (varying potential energies). Familiar examples of this are the rainbow-like patterns formed on the surface of soap bubbles and from the thin films of oil on the surface of roads. This rainbow effect is created by photons first being diffracted, like a prism, then being simultaneously reflected off of the upper & lower surfaces of these transparent films. And when these two sets of reflected photons combine constructively only "one" wavelength of light is seen at a time from a given vantage point from any particular region of the film. The reason for this is that at any given particular spot, and dependent upon the distance & angle to the observer, only "one" wavelength of photons is constructively enhanced by the resulting photon interactions & thus dominant. While all other wavelengths are at some degree of destructive or at least less constructive phase of interference and thus suppressed from being seen by us. Other ways this has been observed is through glass prisms and diffraction gratings.

When interference patterns are produced from only a single wavelength of photons, the resulting patterns consist of bright (photons present) or dark (photons absent) regions or bands.

The result of one particularly interesting interference experiment is one of the cornerstones of what is termed the Anthropic Principle. Which basically states that the observer of any given event affects what they observe - or something like:- that if one observes something one "creates" what they see. From which is concluded that the universe had to create us in order for itself to exist. Something egotistical like this anyhow. Which is basically of less than no interest to myself - except for the fact that no one has had a physical solution for what is occurring during the interference experiment of which we are about to talk of. This interference example, putting it lightly, astounds most people when someone knowledgeable explains the "problem" to them. That is that modern theories would seem to be perplexed by the fact that photons seem not to exist at times & have knowledge of events occurring elsewhere. This may sound crazy, but it's accepted as the truth by many.

The experiment of concern involves the splitting & recombining of a beam of light (a stream of photons) which is simultaneously subjected to interference. Once interference is achieved the portion of the beam on mirror Q side appears enhanced or brighter than before. While inversely it has been completely eliminated from appearing on mirror R. It is at this stage of the experiment that the photons are said to be capable of something which cannot be explained in physical or algebraic terms.

Complete interference is achieved by adjustment of the second beam splitter which lies at the intersection point of the two beams. The venerators, and many others, of the Anthropic Principle state that a reduction in the beams' intensity is analogous to sending out single photons, and are more than just amazed if & when one stops the photons at point A in our diagram resulting in interference stopping & the reappearance of the beam on mirror R. According to them the photons could only reappear on mirror R by somehow knowing a change has taken place. And thus, alter their behavior "willingly". A rather wild interpretation by almost anyone's standards.

Personally, I do remember encountering similar amazement by my own physics teacher & fellow classmates. It puzzled me to - until recently. It was Sid Deutsch who proposed what was happening, and has been supported by work done by Yves Couder and Emmanuel Fort from the University of Paris. Louis de Broglie brought up the idea of pilot waves as a mechanism to account for some of the unusual effects seen in quantum mechanics. Couder and Fort have shown the power of pilot waves with small droplets of silicone suspended over a vibrating fluid mimic quantum effects like that seen in two slit experiments. Sid Deutsch was for me the first who proposed the idea of pilot waves for both electrons and photons for the interference and two-slit experiments. I am not sure if he did not know of Louis de Broglie's work in this area.

I will have to look at the electron equivalence of this experiment, but instead of actual slits the electrons are forced through crystals that act in their place. How do they block a single slit? Or was the original experiment actually done with slits?

Another photon experiment that has made quite a reputation for itself in bringing about a serious exploration into the idea of spooky action at a distance was developed & carried out by a team led by Alain Aspect in the 1980s. This so-called action at a distance or instantaneous communication between two particles, no matter how far apart they are, was based upon a paper (published in 1935) by B. Podolsky, N. Rosen & A. Einstein. Which brought up an apparent contradiction between quantum theory and special relativity. Somehow, or so we are told, information or interaction between two points (the detectors in the experiment) is taking place in a period of time which requires communication by some means beyond the velocity of light. Thus, the contradiction with special relativity. John Gribbin reported in the New Scientist (24th of Nov. 1990) that:- "The key to the experiment is that the polarizations of pairs of photons, produced in <u>a single atomic event</u>, are correlated rather like the spins in the original EPR thought experiment." The key to two photons passing through polarizing filters in opposite directions simultaneously, and being detected to do so, might in fact be a simple physical process since it is supposed to be based upon a single atomic event. If I'm not misinterpreting the data. The photons emitted from a single atom (or molecule?) may well carry with them similar characteristics. Since they were formed within the same event of an individual atom, it seems probable. And that they would also carry the same but mirror momentum characteristics of their component compression & rarefaction while they carry away spin & its orientation components from the atom. Therefore, these photons are of the same nature and thus normally pass through similarly oriented polarizing filters. The universe need not be pre-deterministic - or self-aware as some people would put it. Or having to do with information being conducted beyond the speed of light to explain at least this experiment.

Sid Deutsch's work here was crucial here in revealing a flaw in the logic of Alain's experiments and interpretation of the data. Once again, pilot waves and the twisting of photons through polarizing filters plays a key role. Even if pilot waves did not exist then the non-perfect polarized photons making their odd appearance in the data when they are not expected should not be super-natural but instead an expected outcome of the true dynamic nature of photons and the suspected pilot waves.

The Blueshift & Redshift of Photons

For the longest time a major stumbling block in my FOS Theory series was the fact that I could not account for the aspect of relativity in the blueshift & redshift of light from stellar objects in relation to their varying relativistic motions. Our entire basis of identifying a photon is its energetic value or the work it is capable of doing by affecting in some way the electrons it encounters. The color sensors in our eyes are a perfect example of detectors based solely on the energy delivered by photons, and how the molecules (electrons) interpret them is how we see them.

It is dependent on the direction & the velocity at which the electrons intercept the photons, and also the direction & velocity during the production of the photons that produce these shifts. A photon has a more energetic impact upon an electron if the electron (& it's parent nucleus) is moving towards it. This condition then forces the electron to consume the photon over a shorter period of time & volume of space. While on the other hand if the electron had been basically moving away from the photon then it would be able to consume the photon over a longer period of time (or greater volume of space), and thus the impact upon the electron would be less intense.

One of the best ways to think of these reactions, a photons impact upon an electron, is to think of the various photons as of liquids of varying densities. In our first case though it is only necessary to consider a volume of liquid at one density (a single wavelength of light), and concern ourselves with its motion in relation to some object. For our example, we will consider something to which we can relate to - the interaction between some person and a large volume of water.

If a person encounters a volume of water over a period of a couple of seconds it is perceived by the individual as being "soft", and thus has little impact upon the person. However, if that same volume of water hits the person, or vice versa, in a fraction of a second, then the encounter could leave the person in great pain. As if they had hit concrete. On the other hand, if a person traveling at an Earth-normal fatal velocity were catching up to a body of water at a slightly lower velocity & traveling in the same general direction - then the resulting encounter could make the water seem even softer (or less rigid & dense) than we normally would perceive it to be. Our last example of this type, produces quite the opposite effect. The encounter consisting of the person & the water moving at relatively extremely high velocities towards each other. The resulting effect could be similar to a watermelon striking concrete at a high velocity. Very messy! Photons can have somewhat similar effects upon electrons. That is - in terms of how rigid & dense they are perceived to be by the electrons that encounter them. Or vice versa.

The variation in the density of a photon, which creates the perception of the relativistic principle of blueshift & redshift is directly dependent upon the electrons' motion during the production of a photon. The extra speed of the electrons' parent body (star), in the direction of travel, compresses the accumulating frontal FOS to a somewhat higher rigidity & density than it would normally otherwise be. Enough to be interpreted as a blue-shifted photon by an observer whose platform (or world) is stationary in relation to the photonic source (star). The same interpretation can also take place if the observer is moving towards the photonic source. In this instance, the blueshift (or at least part of it) interpretation is due to the fact that now the electrons involved have less time (or smaller volume of space) in which to consume the photons. Meanwhile, an observer within the source solar system, or within another solar system traveling approximately in the same direction & at a similar velocity, notices no shift as their electrons are traveling rapidly enough to consume the photons over the normal amount of time.

This same solar system which can be observed to produce a blue-shift or no shift is simultaneously also being seen as being red-shifted by those who are moving away from this star or that this star is moving away from. In the case where the star is moving away from the observer, the photons during production are made less compact by the motion of the parent electron. The volume of the photon is enhanced by the fact that during production, the electrons are moving away from the direction of release of the accumulating frontal fabric. Crudely speaking one might be tempted to say that this semi-solid body of the fabric is stretched during its formation. When an observer is moving away from our choice of a star this person's electrons absorbs the photons over a longer period of time or volume of space - due to its general motion away from the photon.

Halton Arp, Quasars & the Redshift of Photons

We have been led to believe that our understanding of the redshift of photons is perfect, and that it clearly shows the expansion of the universe and conversely that there must have been a Big Bang. But one is a fact the other is a theory. The latter cannot be proven it can only be supported by indirect evidence since there are no experiments, within our abilities, that can be performed to show it to be a fact. [And the even distribution of galaxies even when recorded at our current telescopes limits remains uniform, ignoring the groups, and gives no hint of an expansion from some explosion. While our apparent being right at the center of the known universe once again puts us in some privileged position when we used to believe that the Earth itself was at the center of the universe.] The idea of the expansion arose as a result of Edwin Hubble's measurement of the red shift of what he originally thought were simply distant nebulae. Only later did they realize these were other galaxies in the universe. His measurement of the redshift of the galaxies had a direct correlation with their apparent size and brightness and thus distance. [Show graph of galaxies over distance.] However, had he started with looking at the redshift of quasars with respect to brightness the graphs show no linear correlation and thus an expanding universe could not be implied from them. [Show graph of quasar data.]

Halton Arp has collected data that shows us that there is a large number of pairs of quasars whose redshift indicate they are not mirages, or gravitational mirror reflections of each other, and that they appear to be very far away from us - but what is really astounding is that they appear to be associated with galaxies that are much closer to us – based on our interpretation of the galaxies' redshift. These "triplets" appear to be too frequently in close proximity to one another to just be some chance optical alignment phenomenon. Even more astounding is that there appears to be, assuming this is not a thinness of the galaxy as some have suggested, one quasar that is supposed to be far away, but appears to be in front of the galaxy NGC 7319. How can that be? Obviously, it can't be. And thus, obviously there is something wrong with at least one aspect of our understanding of redshift. Or as some suggest, there is something wrong with Halton Arp's data or his interpretation of it. Those who deny all these triplets simply state for this example that: "obviously there is a hole in the galaxy that perfectly lines up with us and thus its light is passing through this hole." Forgetting the unlikely event of a narrow hole passing through the galaxy - the quasar's light is blue-shifted not redshifted. As I previously stated, others claim this is just a thin part of the galaxy, and so the light is simply passing through.

The large redshift exhibited by these quasars with respect to their parent galaxy implies that the redshift is triggered as a result of the high velocity ejection of the material from the galactic core. An extremely energetic event with no equal in energy output. We do know of redshift occurring as light leaves very large and dense stars. This gravitational redshift, of light leaving massive stars, is clearly related to the density changes of the aether surrounding such stars. To me, this would imply then that a quasar-event must trigger a high-velocity in the material, and that it is like the light leaving from a very massive star. Related to relativistic changes as material approaches high-velocities. Thus, we have two different redshifts for the parent galaxy and its quasar offspring. Remember to the blueshift and redshift of the frequency of sound being generated from moving trains. These frequency shifts seem to be of a similar nature.

[More on this in the future second edition of the book.]

Quantum Mechanics Revisited

Now that we have created a near complete model of the physics of … it is time to revisit Quantum Mechanics.

[More on this in the future second edition of the book.]

Quantum Electrodynamics and its' accuracy.

An accuracy of measurement equivalent of measuring the width of North America to the thickness of a human hair.

[More on this in the future second edition of the book.]

Sid Deutsch and the Bell Experiments
[Most content removed until further funding acquired.]

It was engineering educator Sid Deutsch who introduced me to the flaw in Bell's and Alain Aspect's experiments regarding non-locality and the appearance of faster-than-light phenomenon. His ideas regarding this topic are critically important and bring clarity to the flaw in the logic of those who want to believe in the faster-than-light phenomenon. And indirectly once again show the power of Louis de Broglie's pilot waves and the recent [2018] work on pilot waves by Yves Couder and Emmanuel Fort from the University of Paris.

In honor of Sid Deutsch the following is some information from chapter six from his book "Return of the Ether" on the Bell and Alain experiments.

Most of the following from Deutsch has been removed until I can pay for the content. But a small sample of what was to follow is included.

[More on this in the future second edition of the book.]

Twin-State Photon Generator [More to come.]

"In certain experiments involving pairs of photons, to be described below, it appears as if an action visited upon one of the photons is instantaneously felt by the other photon, even if it is relatively far away. John S. Bell pointed out that the correlation between the two photons exceeded the expectation allowed by a local (speed of light) phenomenon [J. S. Bell, 1964, 1987]."

[More on this in the future second edition of the book. After additional funding has been raised to pay for the use of the content.]

The following follows from the work of Sid Deutsch that would have been above this sentence. Once funding is secured then I can pay for the content to appear here.

What follows is based on the content that has been removed:

Two-slit diffraction experiments can be explained mechanically when we consider their influence by de Broglie's pilot waves [and Sid Deutsch's.]. Applying this same influence on photons being able to make it through polarizing filters would provide a mechanical explanation for that which is not supported by pure transverse wave mechanics and today's quantum theory. Once again as per the discussion and data above by Sid Deutsch, the photons pilot waves' reaction to boundary changes is more likely the cause in the required 7.5% distortion needed to destroy the validity of the Bell experiments in Sid Deutsch's conjecture. The only experiments that I can afford to perform, and have performed, are the famous unblocking of light between two polarizing filters that block all light when they are 90 degrees out of phase. By adding a 3rd filter between them at 45 degrees relative to the other two, some light now passes through. Even more, light can be made to pass through by adding additional filters rotated by some other amount, even though the filters in general block out some percentage of any light on their own. There is no pure transverse wave mechanics that allow this. It is more likely that the photons take advantage of their pilot waves motion, due to the displacement of their conducting medium, into the new less dense regions. Just like the spreading of the phonons due to their diffraction around corners, and water transverse-waves through gaps in barriers. Letting people interpret the statistical behavior of Bell's graphs as somehow astounding and magical is not logical or reasonable.

All of the astounding implications of Bell's work is based upon the difference between the Measured versus Expected values in Fig. 6-2(a). All related to a photon count through polarizing filters in alignment and sometimes close enough to be in alignment with these photons still making it through to be counted. Think of what they are trying to imply by something that can be made irrelevant by a photon alignment perturbation of just ±7.5°. Sid Deutsch is not the only one pointing this out, but oddly enough he seems to be forgetting about the influence of pilot waves and their effect on photons and electrons. Even though he was the first person to educate me on them, and then later I learned of de Broglie's work.

Modelling "Spooky Actions at a Distance" by the Animated Physics group:
[More on this in the future second edition of the book.]

Photon Models Not Governed by Bell's Inequality by the Animated Physics group:
[More on this in the future second edition of the book.]

Extending de Broglie's matter wave concept to the Earth, as it rotates and revolves around the Sun, may also point to an additional component in the explanation for both the drift of positive charges to the west, while negative charges drift to the east, and be part of the phenomenon of Frame Dragging. Density is increasing in the direction of travel and decreasing in the opposite direction. Einstein's general relativity gravitation equation model's a masses gravitational field. It also models a density field.

Atoms, Chemistry and Temperature

Cold temperatures on the gas giants allow hydrogen atmospheres to persist/exist and also smaller planets in these similar colder regions of our solar system to retain more volatile compounds that are not held by Mars. Could this be because on Earth, and Mars, the warmer temperatures give the gases the ability to be more buoyant as a byproduct of their kinetic energy and their most weakly held electron's wake dissipation region. Two ideas come to mind. First that on colder planets the hydrogen has electrons are more tightly bound to their nuclei, and thus – we want a reason why they would have less of an opposing gradient for neutral atomic formations, and stay in a stronger gravitational well. Remember that at a purely single-atom level the gas Krypton, $^{36}_{84}Kr$, is more massive than Iron, $^{26}_{56}Fe$, but Krypton is in effect a buoyant gas while Iron appears as a heavy dense metal. It is the charge neutrality of Krypton's outer shell that ultimately allows it to be buoyant. Not its' mass. Why? Why if you cool Krypton enough, to around -155 Celsius degrees, does it become a liquid? The molecular electrical forces and kinetic energy does not drive them apart from the gained kinetic energy of the collisions between them at these temperatures, and the environment they are within because the collisions are just too weak. The Sun appears to repel positively charged atoms because they are too positive with respect to the Sun's aether gradient. So, buoyancy must lie between being too some degree how negatively or positively charged an atom is with respect to its mass. Some of the gases are, in fact, charge neutral while the others must share some common property related to their gradients, along with their hold on their last standard electron or obtained extra electron. Why does hot hydrogen stay around the Sun while heavier but more positively charged ions fly off at increasingly greater temperatures [kinetic energy] the further they move away from the Sun? Is it the electrical currents and magnetic fields that keep hydrogen around like positive ions are trapped in the Earth's magnetic fields? [More on this in the future second edition of the book.]

Gravity Revisited

Gravity is just the outcome of electrons very weakly favoring one side of an atom. In the purest sense "mass" does not bring bodies together, mass attracts electrons, and the electrons bring together masses. Although "mass" in reality is just adherence to the fabric either at a point, or following a path triggered by an induced density anomaly. Einstein's model was based on the observed behavior or extension of Newton's action-at-a-distance field. Well, now we know why. Einstein's equation from General Relativity can be used to model not just a distortion in space-time, portrayed as a distortion in a rubber sheet, but also as the gradient surrounding a mass.

[More on this in the future second edition of the book.]

Explosions – Chemical, Electromagnetic, Mechanical, Nuclear

How and why do explosions occur? This is critical to understanding the nature of gravity and chemical reactions. Whether the explosion originates from a nuclear or chemical reaction ultimately the effect is due to the rearrangement of the electron orbital positions and allowable orbitals around the atoms or molecules in question.

Chemical
[More to be added to explain the rapid rearrangement of electron orbitals.]

Electrical and magnetic
[More to be added to explain the rapid rearrangement of electron orbitals.]

Mechanical and vapor
[More to be added to explain the rapid rearrangement of electron orbitals.]

Nuclear
[More to be added to explain the rapid rearrangement of electron orbitals.]

Stars and Nuclear Engines

Stars constantly bleed off material from them in the form of ions and electrons along with photons. How does the surface of stars and their masses affect the nature of the plasma around them and how they bleed off or eject matter? If we consider gravity as an extension of the positive charge field, the gradient, then positive ions naturally repel away from masses that are large enough. We do see positive charges rise in lightning events. And we do see what appears to be a strange phenomenon of ions speeding away from the Sun and giving thermal energies far greater than the originating ions on the surface of the Sun of, the photosphere, varies between about 6500 K at the bottom and 4000 K at the top (6200 and 3700 degrees C). This temperature actually rises as we transition into the chromosphere and finally into the corona where the temperature apparently exceeds a few million degrees Kelvin. This astounding fact cannot be explained by currently accepted theories, and while such things as magnetic reconnection is being explored by cosmologists and astronomers, who have little training in electrical engineering, they cannot reproduce magnetic reconnections in the laboratory. In the meantime, some engineers are exploring ideas that they know have to have laboratory experimentally verifiable experiments.

[More on this in the future second edition of the book.]

Chapter Seven

THE ELECTRIC UNIVERSE

"Today's cosmology asserts that all cosmic structure resulted from gravitational interactions following a primordial 'Big Bang.' On the contrary, here a Very Large Array (VLA) radio telescope image shows part of the electrical 'circuitry' feeding the core of our galaxy, the Milky Way. No gravitational theorist ever suggested structures of this sort. In electrical terms the red filaments are the cosmic power transmission lines feeding the plasmoid at the center of the galaxy."
Wallace Thornhill

It was when astrophysicists began saying things that I, as an electrical engineer, knew were wrong that I began to have serious doubts about their pronouncements.
Donald E Scott., The Electric Sky

The Electric Universe

Electric Universe Theory (EU) [From their website.]
- Maintains modern astrophysics and cosmology leaves out the Electric aspect of plasma - for example, NASA refers to plasma as "hot gas" in press releases and public offerings
- Says electricity, not gravity, is the defining force of the Universe
- EU has no need of Black Holes, The Big Bang, Dark Matter or Dark Energy to explain the behavior of galaxies
- EU is creating a grassroots revolution in modern cosmology - still academically rejected.

Plasma the 4th known, but most abundant state of matter

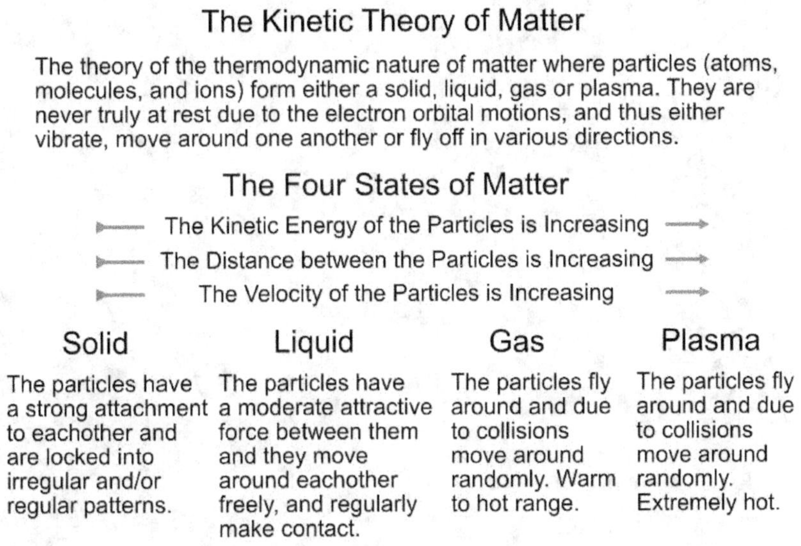

The Kinetic Theory of Matter

The theory of the thermodynamic nature of matter where particles (atoms, molecules, and ions) form either a solid, liquid, gas or plasma. They are never truly at rest due to the electron orbital motions, and thus either vibrate, move around one another or fly off in various directions.

The Four States of Matter

- ▶ The Kinetic Energy of the Particles is Increasing ⟶
- ▶ The Distance between the Particles is Increasing ⟶
- ▶ The Velocity of the Particles is Increasing ⟶

Solid	Liquid	Gas	Plasma
The particles have a strong attachment to eachother and are locked into irregular and/or regular patterns.	The particles have a moderate attractive force between them and they move around eachother freely, and regularly make contact.	The particles fly around and due to collisions move around randomly. Warm to hot range.	The particles fly around and due to collisions move around randomly. Extremely hot.

Figure 86- Plasma, gas, liquids and solids

Plasma flows
Modes of Operation of Plasma flows
1. Dark Current Mode
2. Normal Glow Mode
3. Arc Mode

[More on this in the future second edition of the book.

Electromagnetic Stars & Galaxies

Our galaxy, The Milky Way, appears to in fact be better modeled as a giant electro-magnetic generator. Remember that in comparison to gravity, the strength of electromagnetism is 1×10^{39} times stronger. That is 1 followed by 39 zeros. The proponents of an electric-universe based model claim that it is these giant currents that dominate the galaxies, and give us the observed more uniform motion of stars as galaxies rotate, and not the failed prediction of gravity-based models with that of non-linear motion profiles where there are more slowly moving stars at the outer most edges of a galaxy. Our galaxy has been discovered to be a barred spiral galaxy, and it is this form that Hannes Alfven modeled to be part of an electric circuit that drives the motion of the stars.

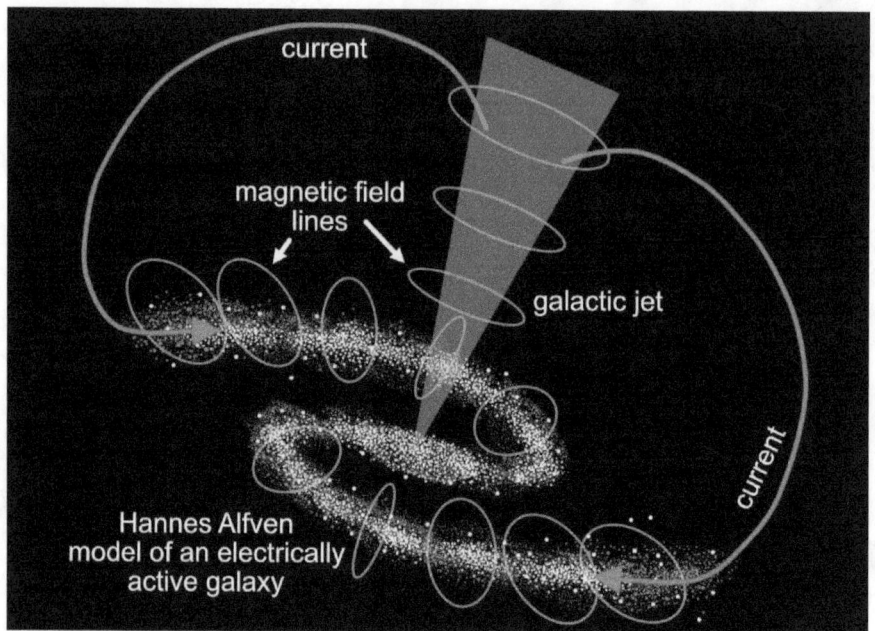

Figure 87- Alfven electric model of a barred spiral galaxy.

Figure 88- Galaxy NGC 1300 Barred Spiral [Credit:NASA]

Figure 89- Galaxies are not just discs in space. Centaurus A (NGC 5128)
Source: ESO/WFI (Optical); MPIfR/ESO/APEX/A.Weiss et al. (Submillimetre); NASA/CXC/CfA/R.Kraft et al. (X-ray)

Experiments and observations show that positive charges are accelerating away from the Sun. There are a few different explanations, and the most obvious explanation is what the EU people are promoting. That is that the Sun appears to act as an anode of positive charge in our solar system with space beyond our Sun acting as the cathode or negative part of a circuit. In the FOS model within this book, we have stated this is because gravity and positive charge are fundamentally the same thing. Varying only in degree and scale. Along with the photons being emitted adding another driving force against the protons and possible additional energy by what the proponents of EU are saying, we end up with more than one factor of why the out layers of the Sun appear hotter, and thus we get faster and faster positively charged ions, when compared to the ions near the surface of the Sun.

What this all implies is that while protons are trying to move apart electrons are trying to bring matter back together.

[More on this in the future second edition of the book.]

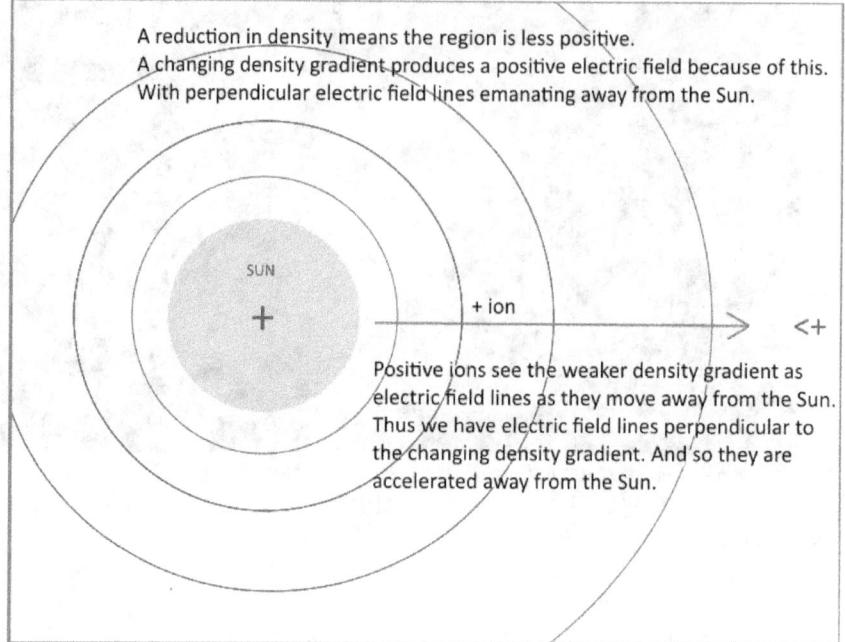

Figure 90- The Sun acts as a cathode ejecting positive nuclei

The Formation of Solar Systems and Galaxies

One of the most common images these days are related to the formation of solar systems throughout the galaxy and questions regarding the habitable zones for which earth like planets might exist. What you rarely here about is the problems with the gravitational based models of solar system formation.

[More on this in the future second edition of the book.]

Strobe Star or Neutron Star? by Ray Gallucci

EU Objections about theorized neutron stars
1. Envision a star, more massive than the sun, spinning so rapidly, yet not flying apart
    a. Astronomers conjured ad hoc the concept of a start so dense that it is composed solely of neutrons packed as dense as an atomic nucleus.

However, a lone neutron decays into a proton, electron and neutrino in about 14 minutes - atom-like collections of two or more should fly apart quite rapidly.

b. Even more damning was the discovery of an x-ray pulsar in Sagittarius with a period of 0.0025 seconds (i.e., rotating at 24,000 rpm, ~ speed of a dental drill).

An even more ad hoc explanation was conjured - this pulsar consisted of matter even denser than neutronium - "strange" matter (perhaps a "quark" star).

c. Further complicating the pulsar theory was the observed varying periodicity of the "glitching" Vela pulsar

It regular speeds up roughly every three years while experiencing "micro-glitches," i.e., random changes in rotation speed. Furthermore, the pulse width also changes with time, sometimes sharply.

This implies that these very massive, unbelievably rapidly rotating stars must instantaneously vary their rates by thousands of rpm.

To ignore the physics required for the propagation of transverse waves because you have no explanation for it is more than just poor science. How good can your scientific method and thus your model be if you ignore data that is inconvenient? This has led us down the path of being unable to understand how the photons could be possibly so high redshifted and our conclusion based on this model is that the speeds must be getting greater and/or the universe is expanding. The alternative to this is simply that longitudinal waves naturally spread out over time and distance. Not because they are losing energy but because they are literally just spreading out. Of course, motion-induced redshift can still be trigged, but these two causes do not need a hypothetical dark energy. So, the whole dark energy requirement is totally bizarre in comparison to photons simply spreading out over time. Which is crazy! Then because of the problem of what is called fine-tuning the anthropomorphic principle is often invoked, or just as unlikely and distasteful is that quantum mechanics is invoked in a manner which they say triggers multiverse formation. With some science [fiction] writers saying that for every point in time where you had to make a choice that in fact both choices were made and two universes came into being because of that. While others have concluded that in most universes, the conditions are not conducive to the existence of life as we know it. And thus, they claim to avoid the fine-tuning problem by saying we just happen to live in the right universe where the conditions are correct. For those who have elected to rejoice in such a formation of multiverses what those who propose it have failed to realize or to tell you is that you would not likely exist at all. They have failed to consider that in such a succession of alternate events the limitations of the biological lottery of your existence. [What is the average life span of a sperm and that of an egg? Sperm reportedly can live up to two to five days inside a woman where they are provided with nutrients from her to help them stay alive as long as possible. With between 30 million and 100 million per event.] Where instead of you pursuing a career and becoming a famous "whatever", the reality is what they could say that might be true for your alternate "you" is also true for your parents, grandparents, etc. backwards in time. So, the likelihood of you existing at all in the multiverse becomes infinitely small in comparison to all the other possible egg-sperm combinations that would be more likely to arise if your parents chose to make love not just on another day, but even just an hour earlier or later. So much for you having multiple lives in other alternate universes. Pure wishful thinking. Or as Conrad Black put it - "When taxed with the extent of the universe and what is beyond it, most atheists now immerse themselves in diaphanous piffle about a multiverse-..." – Conrad Black

Dark Matter versus Electromagnetism

Our galaxy, The Milky Way, is in fact a giant electro-magnetic homo-polar generator. With pairs of quasars as evidence of their association with galaxies and thus showing that electromagnetic dynamics rules galaxies. Fritz Zwicky, in the 1930's, was the first to point out that there apparently was not enough matter to account for the motion of galaxies centered on the newly discovered galactic clusters. However, once the x-ray telescopes as Chandra and XMM-Newton were launched into orbit it was discovered that these galaxy clusters contain large amounts of hot gas, and later shown to contain vast numbers of red dwarf stars, all of which reduced the missing mass/dark matter problem by a factor of somewhere between 10 and 100! Not eliminating it, but showing us that we simply did not have all the facts. Zwicky's 600* times too little mass has been reduced something on the order of between 60 to 6 orders of magnitude. Depending on the clusters in question, and other factors. His work was initially ignored until Vera Rubin discovered something that could easily be measured without any doubt of gravity not working as expected. This evidence for dark matter is supposed to be the flat velocity profile of the stars, first discovered, moving about the center of the Andromeda galaxy. But there is strong evidence that there is no need for this dark matter solution to the flat velocity profile problem as Anthony Peratt has shown that if you simulate galaxies using electromagnetism, rather than gravity, that the flat velocity profile arises naturally. Now part of that flatness may also arise because all of space is a conductor and thus even neutral space farther from the galactic core supports a component of all matters travel through space. Thus, we have not need for the invention of dark matter. The only difference the Electric Universe proponents have failed to consider is that the presence of a mass is the presence of an electric field. They are one and the same and the extreme observed temperature of the Sun's layers may well provide proof of that.

[More on this in the future second edition of the book.]

Chapter Eight

THE COSMIC FOG, REDSHIFT & DEAD STARS

To try to write a grand cosmical drama leads necessarily to myth. To try to let knowledge substitute ignorance in increasingly larger regions of space and time is science.
—Hannes Alfvén (1908-1995)

Space is filled with electrons and flying electric ions of all kinds.
—Kristian Birkeland (1867-1917)

The Cosmic Fog, Redshift & Dead Stars

Our galaxy, The Milky Way, is but a single galaxy among a huge number of other cosmic clouds in the universe. And like the vapor droplets that make up a fog they block some part of the view for someone. With such a huge number of galaxies now known to exist it is easy to envision a cosmic fog, made up of galaxies, around us which we cannot see entirely through or beyond. And even though it is easy to imagine that behind everyone is hidden an untold number of others which we will never see - this is not a visual phenomenon of our view of the universe. If it were then most would think that the sky would be lit up by the billions upon billions of stars that surround us so that there would be no night. Forming a wall of light around us equivalent to the surface of an average star. This does not happen and is commonly referred to as Olber's Paradox.

It is generally accepted that the limiting factor of our view of the universe, and the darkness of night, are the result of the fact that the number of potentially visible stars in the universe is finite, and that the lives of the stars are finite. The fact that stars die is one of the key factors of why our night sky is not lit up like the surface of our sun. If you think about it for a moment it is easy to understand this point. How far in terms of light years can we see? Currently we think it's on the order of about 15 billion light years. Not just a distance but also a time. Yet the average life of an average star is around a hundred million years to less than fifteen billion years. Red dwarfs are estimated to have lives on the order of around hundreds of billion to trillions of years based on our understanding of their energy usage. While our solar system is only roughly 5 billion years. Thus, most stars will have long burned out long ago and stopped filling our view with their light. One of the alternative hypotheses is to account for the darkness of the universe via the fact that most of the light of the universe is red-shifted by the motion of the distant galaxies. Thus, resulting in a loss of visible light, as well as the radio and infrared portions of the electromagnetic spectrum. But some individuals state that the reddening of photons from these ever increasingly distant stars could not be the limiting factor, which prevents the Olber's paradox phenomenon from taking place. Although the reddening of the light due to the motions of the distant galaxies in not considered the cause of Olber's paradox - the expansion, which is the initiator of the reddening, of the universe is considered one of the most important factors (a bit confusing since they are one & the same). Due to the fact that beyond, approximately, 10 billion light-years the stars that lie here are "inadequate radiators over this time scale". In contradiction with themselves it is also stated that "the red shift of starlight from distant galaxies plays an important role in quantitatively explaining the darkness of the night sky.". It is noted that "a successful cosmology must be able to explain Olber's paradox". The FOS Theory series does not have to come up with such an explanation, because it already agrees with the two main accepted explanations. That the death of stars, and the red shift of star light, can account for Olber's paradox. Our vision is limited by and to the photons we can detect. Therefore, any visual phenomenon, or lack of one, is limited to the properties of photons.

The possibility that has gone unexplored, but had been considered for a time, is that photons "decay". Not that they shed energy, as this would be visible, but instead they simply grow in size! Remember that in the FOS series photons are longitudinal waves not transverse waves. And because of this they decay like a sound wave would under ideal conditions. That is the energy contained within the compressed fabric is merely dispersing and so the photon formed will eventually become so spread out, if given the chance that it will revert back to the rigidity & density of that of space itself. It would simply fade away(?). We interpret this lower density which has less of an ability to knock electrons away from atoms as "lost energy." And astronomers have found evidence for this.

This phenomenon can be observed, and is well documented, as far as extremely massive stars are concerned. Here the expansion or decay rate has merely been accelerated by the more intense FOS gradients of these stars making their rate of decay more pronounced than usual. Taking place as the photons leave the fabric gradient (gravitational well) of these stars. In fact, this decay-expansion, and it's inverse, take place within all gradients (gravitational fields) regardless of their size. To some degree or another.

It was long ago discovered that we can detect a red-shift in the photons emitted from and near the surface of very dense stars - where the difference in the FOS layers are very apparent from a microcosmic perspective. In other words, the "gravitational" gradient rises or drops, depending on whether or not your coming or going, very rapidly close in to such stars. From a microcosmic perspective. Another factor besides the structure of a longitudinal wave itself is that what could also be happening is the since the tail edges of an emitted photon, from such a star, are in a slightly more conductive region than the head of the photon. Resulting in the outer edges being conducted away or stretched away from the center of the photon which is in a slightly less conductive region of space. Thus, the photon has its' diameter & wavelength increased. Photon decay has simply proceeded more rapidly.

The inverse has been observed on the earth in experiments that are based on the Gravitational redshift [aka Einstein shift] measured by using the Mössbauer effect.17

It is the vast distances, or periods of time, on the order of billions of light-years that gives these photons time to decay. The photons arriving from the galaxies, at the so-called edge of the universe, have allowed the normal dispersion of these longitudinal waves to become so pronounced that we interpret them as having very large redshifts. The same diagram or graph that shows the rate of redshift of photons from the distant galaxies can also be interpreted as the decay of photons.

What all of this leads up to is that the universe is, according to the FOS Theory series as a result of the longitudinal structure of photons and their natural decay mode, in Steady State. In other words - continually, and forever, in a state of growth, destruction and reorganization. Not the Steady State of the 1950's by Fred Hoyle [et al.] where they tried to fit their model to an expanding universe. This forced them to try and come up with ways of matter being generated out of nothing to fill the increasing vacuum of space as it expands to maintain the same density. Yet, we now hear of modern physicist using spontaneous creation of matter to account for certain effects although they rebalance the effect by having the matter erase itself though another annihilation event to keep the formation to a zero-sum game to not disrupt the universe.

Ironically the so called "final nail in the coffin" for any type of a steady state aether type universe is in fact one of the best supporting pieces of data/evidence for the steady state type of universe proposed by the FOS Theory series. What that piece of data is - is the microwave background radiation that is supposed to be the remnants of the energy of the explosion of the big bang. Discovered in 1965 this radiation of a wavelength of 3.2 centimetres (although now we know it peaks in intensity at one millimetre) has

---

[17] expand and verify this subject - particularly for the blueshift of photons

no apparent source and therefore appears to come from everywhere. Thus, the radiation is termed isotropic. Something is considered isotropic if it is the same when viewed from any direction. [Eric Lerner on its source.] One obvious flaw in the isotropic nature of the CMBR for being evidence of the big bang is that it is still around. We can still "see" them. Microwaves are still photons. Photons travel in straight lines and unless the universe loops back in on itself we should no longer see them. If they are evidence of the big bang then they were released over roughly 13.5 billion years ago [or whatever the current accepted age is.] and have arrived from our visual detectable edge of the universe. Photons generated at the instant of the big bang would start to travel away from the center of the explosion and continue to do so. Traveling ahead of the expansion of matter. Thus, no beings should be able to see the said first light because you would have to be formed ahead of that explosion. Thus, this is a contradiction and a fatal flaw. No matter what point in the explosion of the big bang they occurred at we should not be able to detect them because they should have already passed through and left our space. Out beyond our visible universe.

One possibility, if one accepts the isotropy of these photons, is that these photons have come from all over the universe and in their decay/expansion & diffusion over time the origins of their emissions have long been lost as these photons, from multiple sources, blend together.18 Why this wavelength? More than likely photons (verify) larger (greater wavelengths) than this disappear more rapidly as they are absorbed more readily by the matter spread out in space. While photons of much shorter wavelengths are not yet spread out enough for the source not to be apparent.

Could another explanation exist for this isotropic background radiation? Eric J. Lerner (The Big Bang Never Happened - 1991) reports that he and others have worked out scenarios involving the plasma within magnetic fields dispersing and emitting photons of these very wavelengths. Literally bringing forth this background radiation. Not from some ancient explosion, but from all of the galaxies themselves. (expand on this later)

One of the weakest arguments I've ever heard against any form of a steady state universe is this:- "Also, at Cambridge, Sir Martin Ryle and his team showed that the distribution of galaxies 'out there' is not the same as it is nearer home. Therefore, the universe is not is steady state, and the whole theory is wrong."

So, what if the distribution of galaxies is not the same everywhere! We belong to what is called the Local Group, a small collection of galaxies. And super clusters are now well documented. In fact, it is the existence of these super clusters which are threatening, if not in fact have already done away with, the credibility of the big bang theory(s). For it is the very size of some of these

---

[18]The radio-wavelength photons coming from quasars are so abundant that their origin is apparent. Giving rise to picture quality images.

structures and the length of time necessary for the matter to have moved at normal, and observed, velocities to bring about the existence of these super structures that undermines most big bang theories. Hundreds of billions of years for the formation of these structures vs. roughly some twenty billion years that most big bang theories state as the age of the universe. Even in Stephen Hawking's infinite series of universes each universe is only as long lived as the Hubble expansion indicates. Eric Lerner talks about this in his book - The Big Bang Never Happened – in it he reports that astronomers have discovered groups of galaxies "a billion light-years across; such mammoth clusterings of matter must have taken a hundred billion years to form. Just as early geological theory, which sought to compress the earth's history into a biblical few thousand years crumbled when confronted with the eons needed to build up a mountain range, so the concept of a Big Bang is undermined by the existence of these vast and ancient super clusters of galaxies." These enormous formations also refute a basic premise of the Big Bang - that the universe was, at its origin, perfectly smooth and homogeneous. Some theorists admit that there is no way to get from the perfect universe predicted by the Big Bang model to the actual irregular universe that we live in.

Eric Lerner, from his book, also introduced me to the existence of an electrically based theory which can explain the existence of quasars. Not in terms of black holes, but simply the physics of plasmas, electric currents & magnetic fields on a galactic scale. Basing this theory on plasma research experiments, and in support of Halton Arp's work. Not just on hypothetical math constructions which are the entire basis for the existence of black holes, and for which the limits of validity of the equations are not stated. Thus, when taken to the extreme without the proper limits of applicability, black holes arose from the lack of considering the potential limits of the equations. Engineers know from their first-year classes, that in applying electrical equations without the considerations of a systems limits, that the resulting solutions of using such values leads to potentially infinite power coming out of small systems. Which is not just wrong, but ludicrous and which are obviously never observed. Why because the mathematical infinities merely indicate that maximum current flow is possible. For the current in a circuit as the resistance of the circuit approaches zero, an inappropriate interpretation of the use of this equation would imply infinite current. To simply take the result at face value is a mistake. What the equation implies is that there would be no resistance to the flow of all the electrons in the system. You are not going to get infinite current out of a double A 1.5-volt battery. Instead, what happens is that all the electrons available would be free to travel without resistance. No electrons will materialise out of empty space to provide an infinite supply of electrons. The equations do not prove that infinite power can be obtained from the system. When scientists ignore laboratory experiments in favor of just equations, and thought experiments, this can lead to some extreme predictions that are not founded in reality. Kind of makes one wonder about the state of cosmology today. In particular, when they talk about magnetism in the absence of electrical currents. On a side note, it is very telling of the science that the plasma researchers predicted pairs of "black holes" long before astronomers are claiming to have discovered that there seems to be an unusually high frequency occurrence of pairs of black holes at the cores of galaxies.

The Cosmic Microwave Background Radiation (CMBR) depletion in an expanding universe is rarely ever mentioned. Interestingly enough this is something I had considered some time ago, and it is now being spoken about by some as the eventual depletion of the cosmic microwave background. Not out of absorption, but instead simply due to its passage through space and vacating a region completely. In other words, the CMBR energy should have long ago passed by us and no longer be detectable if it was truly created at the beginning of the universe. The only way it can continue to exist, and not just have passed by us to disappear, is if it were to either be continuously being regenerated by the galactic electrical plasma activity, or as I have proposed from the continuous "decay" (photon redshift by their bodies expanding - "losing focus") of all those distant photons whose source is simply no longer identifiable due to their dispersion.

Why would we still see the CMB radiation if photons travel in straight lines. They should have already moved out of our view and be absent from detection. Unless of course we once again occupy a special place in space and time. Another instance of a human centric universe. Nicolaus Copernicus was the first to point out that the Earth was not at the center of the solar system, and therefore not likely to be at the center of the universe either. One other way for the light to keep appearing, has been proposed, is if the universe falls back, folds back, in on itself. But if that were true then all light and other forms of EM radiation should be falling back into our view. I call this problem the CMBR persistence problem.

The FOS Theory series bases the existence of photons on the structure of longitudinal waves. This wave form is based upon the compression of a fabric of space, which in turn calls for the re-expansion of the fabric & thus conduction of this type of wave. Predicting the eventual dispersion/decay of this wave form. No accepted theory today supports photon decay or tired light as it has been termed in the past, but I cannot avoid it as the FOS theory predicts it. And with it a universe in some form of steady state as a result of its conflict with the premise of the Hubble expansion, and thus for a universe in some form of steady state. One must agree though that no structurally viable form of a steady state theory had been formulated. Until now!

Isn't odd that many scientists, philosophers, and others can accept the possibility of time travel, the multiverse, parallel universes, dark matter and dark energy to fix the theory of the big bang model. But they ridicule the idea of a universe that simply has always existed and shall continue to exist as it does. Why? They may or may not realize it but their big bang theories are based on one simple principle that photons cannot simply decay over time - become redshifted due to decay. Since this would violate the principle of the conservation of energy. Instead, they choose what they call the "simpler" idea that the universe is instead; the product of an explosion/expansion called the big bang, and that the universe may spread out so thinly overtime that no other galaxies will be visible in the future, or that the universe will collapse back in upon itself. Versus the alternative that photons simply spread out which is inappropriately often referred to as decay for longitudinal waves, but true decay would be valid for and required for transverse waves. The physics of spreading out is just part of the natural means of longitudinal wave transmission. A density wave cannot be kept in focus to persist forever at one density. Spreading out means increasing in volume which reduces their density. This is seen as the redshift of photons to astronomers. Which would the principle of Occam's razor favor for the apparent redshift of the light of the most distant galaxies? The entire universe is expanding or... Longitudinal waves spreading out of course.

Chapter Nine

OF TIME & SPACE

Chaos theory only considers the one butterfly's influence, and ignores the other thousand butterflies nearby. – Terrance J. Fidler

Of Time & Space

Time is by far the most elusive of concepts to deal with on almost any basis. This is why I imagine; it's a subject which can be so readily distorted to fulfil the needs & whims of those who wish to convince others of something they've conceived of. Time is a quantity[19] which cannot be contained, and for this reason, can only be "measured" once. But at the same time, it can only be observed indirectly normally as an independent functional parameter of any event. Our view of time becoming malleable arose when the decay of muons arising from cosmic rays seemed to indicate that time could be stretched. This was the proof that time was not constant. But of course, we have proposed is that they decay rate slows down as they feel a back pressure as they enter deeper into the density gradient of the Earth. Without a clear determination of what component muons being generated deeper or at lower altitudes adds to the apparent average half-life.

It is when time no longer appears to be a reliable, and an unchanging independent parameter of events, that our minds tend to draw emotional, and thus wild conclusions about time itself. With some forgetting that the event itself is independent of and has nothing to do with the concept of time or how it is counted. Time is often used to express the limitations and boundaries of events, and it is in this mathematical expressionist role that it has been given unrealistic, or perhaps one should say wishful, properties. It's one thing to use it to express the rate of change in the path of some body of matter encountering

---

[19]We know it exists and it can be measured to some degree but it cannot be confined or run backwards.

the FOS density gradient of a planet, but quite another matter, when a simpler explanation exists, to begin to believe that the time-space coordinates of an event could help someone to go back in time to re-observe the event. This notion, for the most part, has been abandoned by physicists but is still perpetuated by those who wish to believe in time travel or have some way of making a dollar off of the idea in a story. Variations in the readings of atomic clocks, that have been used in the so-called "time experiments", are not proof of time as a fourth dimension that we can go back through to observe events in the past. Even here, physical explanations can be given to account for the variations of atomic clock readings.

To many, the most famous of all the time experiments were those involving "jet-lagged" atomic clocks, while most are unaware of the extended lifetime of muons raining down upon us. Which in reality were a test of gravitational redshift (as opposed to motion-induced redshift) with time dilation as a by-product of this effect. And while Einstein viewed this concept of gravitational redshift as a test of his general theory of relativity, it is considered today as evidence of curved space-time (time as a fourth dimension). Simplistically speaking the experiment involved placing atomic clocks aboard aircraft which were then flown either east or west depending upon the apparent distortion in time one wanted to observe. Sounds so simple, and overall it is, but there's a little more to this type of an experiment than first meets the eye.

Gravitational redshift is a name which actually covers two effects just like the term redshift which can refer to the blue or red shift of photons in relation to motion, but here the red or blue shift of the photons is due to gravity or more precisely the FOS gradient surrounding a mass. We have, in fact already looked at gravitational redshift but referred to its effects upon photons as an example of photon "spreading," not decay. The blueshift component, of a gravitational related shift, induces such a change for the reason that the photon's body is drawn back together because the densest or most conductive region within the FOS is at the head of the photon - thus drawing the rest of its body towards its center via the mechanics of waves. Thereby reducing the photon's wavelength, and thus the photon can be said to be blue-shifted. The same FOS density gradient that induces such effects upon photons also causes the time dilation spoken of in Einstein's special relativity. In fact, the two effects are considered indistinguishable, under the conditions of the experiment we're interested in - for the reason that atomic clocks are based upon a "counter" of a photon signal of a well-defined frequency or narrow range of wavelengths of photons.

The idea behind the experiment is based upon the so-called twin paradox, in relation to special relativity, in which two twins who are separated by one of them taking a trip to another galaxy, at a velocity close to the speed of light, finds that when the traveler returns home its twin, that remained behind on Earth, is much older. This is the consequence of the famous time dilation effect, in which the traveler's clock slowed down relative to the twins who stayed at home. The clock slowing down as a result of it approaching the speed of light. Whose mass would also increase an incredible amount, but they don't usually mention that.

The experiment is actually a little more complex than it first sounds. The reason for this is that there are two effects taking place at once, and because of this, one must consider both of them in their calculations for determining the relative consequence of each effect. For the reason that they can offset each other and therefore confuse the observer if one forgets to take this into consideration. The experiment, as I know of it, forces one to not just compare the two clocks to one another, but in fact to compare the two of them with that of an imaginary clock that is theoretically stationary with respect to the center of the Earth. Because a clock stationary at some point on the surface of the Earth is in fact moving at quite a velocity in revolution about the center of it, and therefore is experiencing both time dilation as a result of being in motion, and blue-shift for being at some distance from the center of the planet. It's because of these effects that Einstein clearly stated in his theory of special relativity that a moving clock must always be compared to a clock that is in an inertial frame or in other words at rest or moving in a straight line at a constant velocity.

The results of the experiments showed that generally speaking if one flew a clock to the east that this clock would tick more slowly than both a clock on the surface of the Earth and even more slowly with respect to the clock in the inertial (imaginary) frame of reference. The clock on the surface ticks more slowly than the one in the inertial frame because it too is in circular motion about the center of the Earth, and thus is experiencing time dilation (not traveling back in time - just lagging behind in time or more precisely behind in counting). But the clock on board the aircraft is also experiencing blue-shift and depending on its altitude this can offset the time dilation its experiencing or at least its counting. In fact, if its velocity is low enough and its altitude high enough then the ground clock will be ticking more slowly than the airborne one. Even though the clock that is flying is traveling at a greater velocity it's time dilation is offset by the greater blue-shift (as opposed to the ground clock's blue-shift) induced by it's greater altitude. On the other hand, it was found that a clock sent off to the west always ended up ticking more rapidly (Traveling forwards in time? No! Just ticking faster.) than the surface clock. In this case, the gravitational blue-shift due to the altitude of the aircraft helps increase the clock's count or rate of travel into the "future" by some peoples interpretation. Though if your new to these facts about such experiments I'd advise you not to get too excited about the "apparent" implications of such experiments.

When the experiment was first performed in October of 1971, by a U.S. team of scientists led by J.C.Hafele and Richard Keating - they found for the eastbound flight a lag in time of only 59 nanoseconds (billionths of a second). However, they had predicted a lag of 40 nanoseconds. The time dilation would have been larger but it was reduced by the blue-shift of the aircraft being in flight and therefore at some greater altitude than the normal distance from the center of the planet. For the westbound flight, the blueshift helped the time effect along which was taking a small leap forward anyhow. In this instance the blueshift accounted for approximately two thirds of the jump in time which came to be 273 nanoseconds. Two nanoseconds short of the predicted time. The discrepancies in the predicted times and actual times being accounted for by experimental errors due to the inaccuracies of the flight data and variations in the rates of the clocks due to their structural limitations and functional characteristics. It is because of these incredibly small values of distortions in time at these velocities that one often hears that velocities approaching the speed of light are necessary to experience worthwhile time dilation - like those that take place in the movies. And would be possible if it were not for the fact that presently, and more than likely very far into the future, it is close to impossible to get anywhere near the velocity of light. Simply because of the enormous amounts of energy necessary to do so. Or so it is believed by most people who think that time travel is possible or at least not impossible. Again, we are not even considering the increase in mass as a body approaches the speed of light.

One major problem though is that I've already come up with a physical solution for three of the time variation phenomena. Those are the altitude based blueshift (that has nothing to do with velocity), particle longevity at high velocities, and for the more rapid ticking of west bound flights. So, if I can account for them now it seems likely that the FOS theory series will also be responsible for physical solutions for the others in time. Leaving time travel without a foot in reality.

[Could we have equations relating the time dilation to the resonant frequency changing with a changing mass of the systems in question? More on this in the future second edition of the book.]

Although the effect of time dilation is real and so are its effects upon photons and atomic resonations, in such a manner to give the perception of time variations, it is unlikely to induce such effects that are dreamed of in science fiction stories. I say unlikely because I'm not sure, but I do know that with three of these effects accounted for by the FOS model that it's just a matter of time for the others. Solutions not related to any additional dimensions of space or it's being warped. And if the warping of space or time as a fourth dimension are the only acceptable possible causes of these distortions, then I'd be tempted to say these ideas are wrong. And so is the idea of time travel.

The physical basis of the so-called time dilation effects, based upon the altitude of a body, is that at any given altitude the wavelengths or frequencies of oscillation of the atoms within an atomic clock change due to the relative FOS density surrounding the atom. The lower the density outside the atom the closer to the nucleus the electrons should move, tightening them around it, and at the same time increasing the frequency of oscillation. Thus, with a frequency shift time appears to be passing more quickly. Time is not changing, just the frequency of oscillation.

The unit of time we call the second, as defined in 1967 by the International System of Units (SI), is the duration of 9,192,631,770 cycles of radiation corresponding to the transition between two energy levels of the ground state of the caesium133 atom. This definition makes the cesium oscillator an atomic clock, and the primary standard for time and frequency measurements.

The Special and General Theories of Relativity

What perhaps we should be really talking about for clarification is Galileo Galilei, Isaac Newton and Albert Einstein's Special Theory of Relativity. While the General Theory of Relativity was an extension to cover objects in any form of motion and Einstein's model of gravity as a deformation in the space time continuum. [More on this.]

The Special Theory of Relativity's original name was "On the Electrodynamics of Moving Bodies"

What is the Special Theory of Relativity? Einstein's summary of Special Relativity is simply this:
1. The laws of physics are the same for all observer's in uniform motion.
2. The speed of light is a constant.

General Relativity can be summed as:
1. The laws of physics are the same for all observer's regardless of their motion.
2. Inertial mass and gravitational mass are one and the same. And this is called the Principle of Equivalence.

Einstein developed a mathematical representation of a gradient that has been represented by a deformed sheet. This has been our representation for over a century of how gravity distorts space & time. The greater the mass the greater the deformity. Or so the theory goes. A gradient can be formed several ways. One could be just the residual positive charge left over that the electrons could not neutralize. Remember that the difference in strength between gravity and charge is 1 followed by 39 zeros in magnitude of difference.

[More on this in future second edition of the book.]

CONCLUSION

The surest way to corrupt a youth is to instruct him to hold in higher esteem those who think alike than those who think differently.
Friedrich W. Nietzsche (1844 - 1900)

In conclusion, we have tried to show that using a longitudinal form for photons, which consists of a compression and rarefaction cycle of transmission in the structure itself of such a wave, we get the gradient form that is an electric field, and that positrons/protons appear to be standing waves. Electrons arising from two-photon physics with their wake dissipation regions trigger the formation of quantized orbitals around nuclei, and act to hold nuclei together. Thus giving us both nuclear physics and chemistry. The positrons/protons creating a permanent density gradient around them via the compaction mechanism inherent in a standing waveform. And so, as to the secondary title of the book, the relationship between gravity and quantum mechanics, the only difference between gravity and quantum fields is that gravity is a large scale structured gradient field whereas quantum fields are just pockets of density gradients within atoms, and in some cases outside of atoms, that strongly attract electrons when compared to the lower density gravitational gradients. With all this being said we see that this implies a Theory of Everything.

I hope you can bear with me, and support this work, as I attempt to gain additional funding to be able to pay for additional content that is from the work of others. As both supporting evidence, and in general that I am not the only one seeking out a physical explanation for all that is around us.

I have created a website to both promote the work, and to try and find some additional funding for having it translated into other languages. The website is thedeathofthedarkenergyidea.com. Here people can contact me directly, and also find out more information on the development or sharing of artwork, animation and simulations for the model. I am also hoping to find a professional editor.

ABOUT THE AUTHOR

I have a diploma as a Robotics and Automation Technologist, and a degree in Computer Engineering – specializing in hardware. However, due to some bankruptcies of the companies I had worked for, and layoffs I am not working directly in this field. Beside some minor work with the RaspberryPi mini-computer, I have not been working much on anything related to robotics. Besides working on this book and the concepts related to it, I don't really have the space where I live, nor the funds to do what I want to do in farm robotics. Let alone have the funds to purchase and run a farm efficiently, and have a group of people work on farm robots.

My favorite hobbies are/was scuba diving and underwater photography.

I took up robotics because I was interested in robotics to reduce the use of pesticides on farms, sort through garbage, and to develop automated cameras for underwater photography and monitoring dive sites. Using this technology, parks and game reserves could monitor the health and wellbeing of these places, and everything that lives within them. A normal drone cannot be used due to their limited power supply, and because of this most have short flight times of usually less than 30 minutes. Whereas a solar powered automated system that does not move, or moves very little could work and communicate for weeks, months, or longer.

PRIMARY PATRONS & SPONSORS

A list of some of my Patrons & Sponsors:
(Go to the main web site to sign up to become a Patron. Note due to limitations for the printed editions. Only one printed page per level or category, and by date, will show for Patrons & Sponsors.)

Anonymous Amount:
1. J.D., Chicago, Il, USA, Level-Anonymous, Keep up the fight for reason.
2. Robert F., Toronto, Ontario, Canada, Level-Anonymous, Dedicated to K.F.
3. J.P., Boston, MA, USA, Level-Anonymous, In Loving Memory of Felix. You brought us so much joy and love.
4. Hank & Mary Funk., BC Canada

Bronze

Silver

Gold

Platinum

Unobtainium

Time based:
1. J.D., Chicago, Il, USA, Level-Anonymous, Keep up the fight for reason.
2. Robert F., Toronto, Ontario, Canada, Level-Anonymous, Dedicated to K.F.
3. J.P., Boston, MA, USA, Level-Anonymous, In Loving Memory of Felix. You brought us so much joy and love.
4. Hank & Mary Funk., BC Canada

People who provided useful feedback that was used:

A list of some of those who provided helpful feedback that influenced some editing, and additions to the book.

Jack Musser, USA

INDEX

action-at-a-distance, 34
aether, 34
angular momentum of photons, 141
annihilation reactions, 132
Anthony Peratt, 30
antimatter-matter annihilation, 35
antiproton, 133
Aufbau principle, 146
black body radiation, 163
Bohr ionization potential, 126
Breakwater gap diffraction, 19
breakwater gap diffraction,, 17
Classical Electron Radius, 114, 116
Classical Theory of Light, 174
CMBR persistence problem, 25, 204
color confinement, 127
Cosmic Microwave Background Radiation, 7
CRC Handbook of Chemistry, 115
dark energy, 36
Dark Energy, 7
density gradient, 37
Doppler Effect, 4
electrical force constant, 141
Electron Capture, 114
Electron Orbital Capture, 114
Electron Wiggler Laser, 169
Emmanuel Fort, 183
epicycle model, 32
Fundamental Model, 138
Fundamentals In Nuclear Physics, 122
general theory of relativity, 37
Hans Mes, 119
Hideki Yukawa, 113
Higgs field, vi, 8, 9, 22, 34, 37, 38, 42, 51, 61, 66, 81, 84, 85, 86
Higgs mechanism, 22
Higgs particle, 22
Horizon problem, 7
Interference patterns, 182
James Clerk Maxwell, 166
Johannes Kepler, 15, 85
Louis de Broglie, 21, 175
magnetic force constant, 141
Michael Faraday, 166

Michelson & Morley, 33
neutrinos, 132
neutronic system, 122
Nuclear Potential Barrier, 124
Occam's razor, 119
Ohm's Law, 32
photoelectric effect, 171
pilot waves, 175
pions, 118
Pions, 113
plasmoid, 109
polarization, 28
polarizing filters, 28, 39
poloidal and toroidal rotations, 75
quantum effects, 39
Quantum Tunneling, 180
Quarks, 129
Richard A. Beth, 140
Rutherford, 24
Sir Isaac Newton, 85
Space-time, 37
Standard Model, 129
Steady State, 201
String theory, 41
transverse waves, 39
Two gamma-ray physics, 40
Two-Photon physics, 35
Ultra-cold neutrons, 120
UV Catastrophe, 164
wake dissipation regions, 171
Wake dissipation regions, 17
Yves Couder, 16, 183

LIST OF FIGURES

Figure 1- c+v for the Michelson and Morley experiments 3
Figure 2 Michelson and Morley experimental setup 4
Figure 3 - The Doppler effect for sound 5
Figure 4- The Doppler effect for light [Credit: NASA image of M31] 5
Figure 5- Hydrogen lines of the Balmer electron orbital transition series 6
Figure 6 - return to the equilibrium position requires a restoring force 9
Figure 7-Transverse wave mapped onto a longitudinal wave 14
Figure 8 - circular transverse waves spreading out on a pool of water 15
Figure 9 Particle compression is Not 100% directly forward, this triggers redshift 15
Figure 10- Two slit experiments for electrons or light 17
Figure 11- As the gap distance approaches the wavelength spreading maximizes 18
Figure 12-Breakwater gap diffraction of water waves 19
Figure 13- Boundary wave generation 19
Figure 14- Diffraction of particles compared to waves 21
Figure 15- Diffraction of low frequency vs high frequency sound waves and photons 22
Figure 16- Single slit diffraction of photons ... 22
Figure 17- Higgs mechanism versus mass by standing waves 24
Figure 18- Higgs boson - the "cocktail party" analogy 24
Figure 19- CMBR some 380,000 years after the Big Bang to CMBR void formation 32
Figure 20 - transverse wave and longitudinal wave polarizations 34

Figure 21- Density gradient profile equivalent to distortion in space-time................................. 44
Figure 22- Standing wave with non-point like scattering 53
Figure 23- Distortion in space-time depicted by balls on a rubber sheet.. 60
Figure 24- Both transverse waves and longitudinal waves can be polarizable 61
Figure 25 The more polarizing filters the more light passes through....................................... 63
Figure 26 Pilot wave reflection near an edge leads to a higher density plane. 64
Figure 27 Pilot waves induce both photons and electrons to align with the more conductive path. 64
Figure 28- Transverse wave amplitude reduction 73
Figure 29- Inverse Square law for sound wave intensity reduction 74
Figure 30- A transverse wave form mapped from the components of a longitudinal wave 79
Figure 31- Two-photon physics giving rise to an electron-positron pair................................... 80
Figure 32- Electron generates a wake dissipation region as it passes through space................... 83
Figure 33- Atomic and cloud charge distribution within the Earth's gravitational influence 84
Figure 34- Standing wave points in space 88
Figure 35- Mass gradient versus an atom's quantized gradient tiered structure 89
Figure 36- An atom's electron generated wake dissipation regions [exaggerated] 95
Figure 37-probability map of the location of a particle from a simulation using GNU Octave where its path is fluctuating due to particles affect from its last orbit 96
Figure 38- The derived charge and magnetic moment densities of the proton and neutron 98
Figure 39- Electronegativities of the elements 106

Figure 40- Electron affinity of the elements . 106
Figure 41- The more weakly held electrons are more sensitive to FOS gradients 108
Figure 42- Atomic and cloud charge distribution within the Earth's gravitational influence 109
Figure 43- The Sun appears as a more positive body to positively charged ions 110
Figure 44- Strong and weak principles of equivalence... 112
Figure 45 - weakly-held electron distribution changes 113
Figure 46 - the more weakly held electrons are more sensitive to FOS gradients 113
Figure 47 - electron motions with respect to the strong principle of equivalence 114
Figure 48- 2019 astronomical image of a black hole or electrically based plasmoid? 116
Figure 49- Wake dissipation action by electrons creates quantized regions........................... 119
Figure 50- a negative charge profile around neutrons... 129
Figure 51- The Standard model 143
Figure 52- The Standard model families or hierarchy.. 144
Figure 53- The Standard model with SUSY, supersymmetry and sparticles 144
Figure 54- The refined Standard model or Fundamental model 145
Figure 55 Electromagnet and Permanent Magnet and their electron motions......................... 149
Figure 56- conventional current vs electron current... 151
Figure 57- Ar, V, Fe and Cu ionization energies in eVolts 152
Figure 58-An Aufbau diagram for Nickel...... 153
Figure 59-Electron orbital energies in eV for Fe, Co and Ni from the CRC Handbook 153
Figure 60-positive, neutral & negative particles moving thru a magnetic field 154

Figure 61- voltage driven electron flow along a wire between to magnets 154
Figure 62- electron induced aether flows around a circular path 156
Figure 63- Magnus effect and its electron equivalent 157
Figure 64-atomic magnet balanced and unbalanced systems 157
Figure 65- electrons are not passive 158
Figure 66- Positive ions are more passive 159
Figure 67- halo region created by single planar orientation 160
Figure 68- atomic magnet changing orientation with external magnetic field 161
Figure 69-electrons traveling in the same direction attract each other by increased denser regions being created by the opposing magnetic fields and the Magnus effect 162
Figure 70- electron current flow in two wires in the same direction 163
Figure 71- electron current flow in two wires in opposite directions 163
Figure 72- right-hand motor rule 164
Figure 73- left-hand generator rule 165
Figure 74-electrons induced to move along a wire 165
Figure 75- Proton drift vs electron drift with mirror points 166
Figure 76- Magnetic field B and charged particle interaction around a wire 167
Figure 77- typical representation of an electromagnetic wave 172
Figure 78- Transverse wave form profile from a longitudinal wave body 173
Figure 79- a compression wave is equivalent to a gradient body 173
Figure 80- Atomic gradient is an electric field 174
Figure 81- wake dissipation regions around an atom. Exaggerated scaling. 177

Figure 82- Lyman, Balmer and Paschen series for Hydrogen.. 179
Figure 83- Reflection and Diffraction around Breakwaters 182
Figure 84- Sound wave and light wave single slit diffraction 183
Figure 85-Quantum tunneling and wake dissipation effects.. 187
Figure 86- Plasma, gas, liquids and solids ... 199
Figure 87- Alfven electric model of a barred spiral galaxy.. 200
Figure 88- Galaxy NGC 1300 Barred Spiral [Credit:NASA]... 200
Figure 89- Galaxies are not just discs in space. Centaurus A (NGC 5128) 201
Figure 90- The Sun acts as a cathode ejecting positive nuclei... 202

LIST OF TABLES

Table 1- Transverse waves versus Longitudinal density waves. 16
Table 2 A list of the major fields in modern physics 27
Table 3- Nine different protons formed from different quarks [2 ups, 1 down] 50
Table 4- Nine different neutrons formed from different quarks [2 downs, 1 up] 51
Table 5 - Proton radius compared to electrons, muons, pions 85
Table 6- atomic data of the three main hydrogen isotopes 103
Table 7- molecular properties of the first two hydrogen isotopes 103
Table 8- physical properties of water made from the three main isotopes of hydrogen 104
Table 9- Isotopes that can experience Electron Capture 121

Table 10 - Proton radius compared to electrons, muons, pions 122
Table 11-electron, muon, pion and tau comparison table 124
Table 12 - Nine versions of protons formed from different quarks [2 ups, 1 down] 136
Table 13 - Nine versions of neutrons formed from different quarks [2 downs, 1 up] 136
Table 14- Some physical properties of ordinary and heavy water (source lost) 185
Table 15- Some comparative properties of hydrogen versus deuterium (source lost) 185
Table 16 - Other physical differences between hydrogen and deuterium (source lost) 185

Bibliography

Arp, Halton C. 1998. *Seeing Red.*
Baggot, Jim. 2013. *Farewell to Reality: How Modern Physics Has Betrayed the Search for Scientific Truth.* Pegasus Books.
Deutsch, Sid. 1999. *Return of the Ether: When Theory and Reality Collide.* SciTech Publishing.
Lerner, Eric J. 1991. *The Big Bang Never Happened.* Vintage Books, a division of Random House.
Sagan, Carl E. 1985. *Cosmos.*
Scott, Donald E. 2006. *The Electric Sky.* Mikamar Publishing.
Smolin, Lee. 2006. *The Trouble with Physics.* Houghton Mifflin Harcourt.
Unzicker, Alexander. 2013. *The Higgs Fake.*
Woit, Peter. 2006. *Not Even Wrong: The Failure of String Theory and the Search for Unity in Physical Law.* Basic Books.

Additional Information

Adobe Photoshop Elements 14
Affinity Designer
GNU Octave is a high-level language, used for numerical computations and simulations. With a high degree of compatibility with Matlab. It is also FREE, and has a great many resources on the World Wide Web to help you use it successfully for your computation intensive projects. Thanks to everyone who provided examples I could use.
Grammarly
Mathematica by Wolfram was used to generate a number of the images. Along with some of them being generated by using the Wolfram Alpha browser app. Thanks to those who provided examples of various functions on the community sites.
Paint – mostly used to ensure bitmap image compatibility for inserting images into MS Word. Using a bitmap is more efficient because of the way MS Word usings bmp images.
Paint.net was used initially to create and edit a number of the images. But then I found, and started using the vector-based drawing program Affinity Designer.
SnagIt, by Techsmith, was used to edit some of the images.

Sources of information and Literature

Encyclopedia Britannica
Eric J Lerner: Lppfusion.com

Wikipedia

Literature

Arp, Halton. Seeing Red. Montreal, QC: Apeiron, 1998.
Baggott, Jim. Farewell to Reality. Berkeley, CA: Pegasus Books, 2013.
[A work in progress.]

World Wide Web sources:

Commons.wikimedia.org
ESA.int
MCAT-Review.org
NASA.gov
Physcis4spm.com
QuoteAddicts.com
Sdsu-physics.org

And many other sources for inspirational images, images containing technical information, novel ideas, and others. I will be adding to this list over time.

www.ingramcontent.com/pod-product-compliance
Lightning Source LLC
Chambersburg PA
CBHW072027230526
45466CB00020B/1012